SpringerBriefs in Applied Sciences and Technology

For further volumes:
http://www.springer.com/series/8884

Melvin Choon Giap Lim
ZhaoWei Zhong

Carbon Nanotubes as Nanodelivery Systems

An Insight Through Molecular Dynamics Simulations

 Springer

Melvin Choon Giap Lim
School of Mathematics and Science
Singapore Polytechnic
Singapore
Singapore

ZhaoWei Zhong
School of Mechanical and Aerospace
  Engineering
Nanyang Technological University
Singapore
Singapore

ISSN 2191-530X   ISSN 2191-5318   (electronic)
ISBN 978-981-4451-38-3   ISBN 978-981-4451-39-0   (eBook)
DOI 10.1007/978-981-4451-39-0
Springer Singapore Heidelberg New York Dordrecht London

Library of Congress Control Number: 2013934990

Printed on acid-free paper

Springer is part of Springer Science+Business Media (www.springer.com)

Preface

This book describes the usage of a carbon nanotube channel as a nanodelivery system for copper atoms, in which the transportation of copper atoms is achieved through the application of electromigration. Apart from that, it also highlights the use of molecular dynamics simulations as a means for the investigation of the nanosystem. The steps involved in building the molecular dynamics simulation program is illustrated in this book, leading to an example showing the application of the molecular dynamics program, and the use of the program to investigate the carbon nanotube as a delivery system.

This is a text intended for engineering and science students, who wish to create a molecular dynamics simulation program of their own and to conduct an analytical study of a molecular system. The nanodelivery system of carbon nanotubes presents the possible usage of the carbon structure in many areas in the future. The reason for writing this particular subject is to share the expertise of this field of study, which is relevant to current development in nanotechnology.

The book is divided into two main parts. The first part introduces some of the relevant concept of molecular dynamics simulations, and to show how to use it to study a group of atoms, which are formed into a rectangular system. The second part focuses on the application and results of the molecular dynamics simulation of carbon nanotube systems, and at the same time provides scientific explanations to the phenomena observed during the simulation process of the carbon nanotube as a nanodelivery system.

This book is comprehensive, informative and serves as a guide for readers who are interested in molecular dynamics and the latest development of carbon nanotubes. The content of this book consists of a research effort that lasted for 4 years, and inspirations from many well-established scientific works.

Lastly, we would like to thank our friends and families for their constant support and understanding towards our busy schedules.

M. C. G. Lim
Z. W. Zhong

Contents

Abbreviations

CNT	Carbon nanotube
COM	Centre of mass
EAM	Embedded atom method
LJ	Lennard-Jones
MD	Molecular dynamics
MPCVD	Microwave plasma-assisted chemical vapour deposition
MWCNT	Multi-walled carbon nanotube
SWCNT	Single-walled carbon nanotube

Symbols

Roman Letters

a	Acceleration
a_0	Lattice parameter
C	Atom-type dependent constants
d_c	Diffusion coefficient
e	Electron charge
E	Energy
F_d	Direct force
F_w	wind force
k_m	Temperature-independent quantity for electromigration
K	Reciprocal lattice vibration vector
k_b	Boltzmann constant
m	Mass of an atom
N_l	The number of atoms in the lth atomic layer
N	A fixed number of atoms
P	Pressure
r	Distance between two atoms
$R(T)$	Resistivity
r_v	Radius vector
s_l	Layer structure factor
s_p	Atomic position
t	Time
T	Temperature
u, v, w	Velocities
V	Volume

Greek Letters

γ	Force on an atom
β_f	Electric field
μ	A fixed chemical potential

ξ	Embedding function
γ_{em}	Electromigration force on an atom
ς	Host electron density
$\Psi(r_{ij})$	Two body potential

Chapter 1
Introduction

1.1 Background

A carbon nanotube (CNT) is a fascinating nanostructure that has promising potentials for future applications. A CNT is a cylindrical tube made of carbon atoms, and exists either as a single-walled structure known as a single wall carbon nanotube (SWCNT), or as a multilayered structure known as a multi-walled carbon nanotube (MWCNT). The diameter of an SWCNT can be as small as 0.3 nm [1], whereas the inner diameter of an MWCNT can be larger than 15 nm [2]. An article, which first reported the findings of CNTs, was published by Radushkevich and Lukyanovich [3], in the Russian Journal of Physical Chemistry. However, the most significant publication of CNTs made to the scientific community was given by Iijima in 1991 [4]. Intensive research efforts on CNTs have since escalated and gained attentions worldwide.

There are a wide range of applications in which a CNT can be used. They include implantable materials and devices, surgical aids [5], DNA detectors and sensors, molecular transporters [6–11], and many more [12, 13]. A CNT can also be combined with other biomolecules to form a hybrid system for bioelectronics and nanocircuitry [14]. Since the centre core of a CNT is hollow, a CNT is naturally a good candidate as a nanochannel. The studies involving the flow of molecules and atoms along a CNT are thus essential for investigating the suitability of the CNT as a transport vehicle on a nanoscale, and predicting the flow phenomena that can be expected in the nanochannel.

There has been a particular interest in SWCNTs as nanochannels because the wall of a SWCNT is only one atom thick, and the atomic influences of the carbon wall on the flow substances are less complicated compared to a MWCNT, where the multilayered wall effect may need to be considered. However, the structure of an SWCNT is less than perfect. SWCNTs have been produced experimentally over the years [15, 16] and often, defects can be found along the walls of SWCNTs [17, 18].

In usual cases, a SWCNT can be produced with well-defined morphologies, a uniform thickness and other unique characteristics, which allow the structure to be

M. C. G. Lim and Z. W. Zhong, *Carbon Nanotubes as Nanodelivery Systems*,
SpringerBriefs in Applied Sciences and Technology,
DOI: 10.1007/978-981-4451-39-0_1, © The Author(s) 2013

made suitable for encapsulating and transporting metal atoms. The properties of CNTs can be described using a set of vectors, known as the chiral vectors, to determine the arrangement of the carbon atoms, the type of electronic structures and also the diameter of the CNT [19]. The electrical conduction in an CNT can either be metallic, or semiconducting, depending on the indices in the chiral vector of the tube [20]. Other additional features of a CNT include kinks and nanotube junctions, which are usually formed by combining more than two CNTs together [21, 22].

A pentagon-heptagon pair is an example of a defect that exists in the network of the CNT [23]. In fact, a pentagon-heptagon pair defect can occur when two CNTs of dissimilar diameters are connected together experimentally [24]. A junction is formed when two dissimilar CNTs are joined together at their ends. The structure of the junction is similar to a nozzle or a diffuser, depending on how the cross sectional area of the junction changes along the CNT channel [25]. The shape of the CNT channel junction affects the volume of the encapsulated material flowing through it. The junction of the CNT channel is therefore a crucial factor to be investigated for the flow properties of nanomaterials [26].

The filling and encapsulation of metal atoms in CNTs is a widely discussed topic because there are many properties of CNTs that can be enhanced through the addition of metal atoms in the cavities [27, 28]. Besides that, CNTs can also be used as templates for forming nanowires [29, 30]. For example, iron filled CNTs are suitable to be used as magnetic field sensors due to the ferromagnetic behaviour of the system at room temperatures [31]. Apart from that, CNTs filled with ferromagnetic fillers can also potentially be used to control the heating of tumour tissues [32].

The studies involving the filling of materials in CNTs have been carried out for many years [33–36]. There are many methods, which have been explored and experimented to fill metals into CNTs. Some of these methods are capillarity-induced filling, carbon arc discharge methods, wet chemical techniques, pyrolysis, and many others [37]. Quite often, simulation methods are also attempted by researchers to investigate the possible filling processes of metal atoms in CNTs [38, 39]. One of the methods of filling copper atoms into CNTs is through the sucking effect caused by the capillary attraction between copper atoms and the CNTs, achieved through a process known as the microwave plasma-assisted chemical vapour deposition (MPCVD) method [40]. By using the same principle, the capillary effect of palladium in CNTs has been tested using molecular simulation techniques [41, 42].

Since CNTs are suitable for encapsulating metal atoms, CNTs are thus ideal candidates to function as delivery systems for nanomaterials, such as metal atoms. There are several ways of transporting encapsulated materials in CNTs. Electromigration is one of the important methods for transporting metal atoms along CNTs [43]. The forces that contribute to the electromigration force depend on the electric field and the scattering of electrons, which are referred to as the 'direct' and 'wind' forces, respectively [44]. By altering the electric field along a CNT, the migration of the encapsulated metal atoms can be achieved due to the effect of electromigration. The encapsulated metal atoms move towards the opposing direction of the electric field that is parallel to the direction of the electron flow. Such phenomenon is observed in experiments and simulation studies [45–47]. Besides that, the voltage gradient determines the direction of the mass transport along CNTs [48].

In the case of atomic scale mass transport, we can control the mass distribution along CNTs, since the direction and the rate of the mass transport are affected by the external drive. The importance of such a method is that the mass delivery of metal atoms in CNTs can be applied to nanorobotic spot welding using electro-migration processes [49]. The nanorobotic manipulation is carried out to transport copper mass through inter-nanotubes. The magnitude of the bias voltage to transport copper along CNTs, with diameters ranging from 40 to 80 nm, is between 1.5 and 2.5 V.

Quite often, the studies involving the transportation of metal atoms along CNTs are carried out by considering uniform tubules as the delivery systems [50]. Non-uniform tubules, on the other hand, are rarely looked into. For example, the feature of a CNT channel junction is worth investigating, because the structural geometry of the junction affects the outcome of the flow processes along a CNT. The investigation of flow processes along these CNT channels provides valuable information, which could be useful for manipulating the outcome of the mass flow using the feature of a channel junction.

In view of the complexity of the experiments for the studies of CNT channels as nanodelivery systems, we investigate the nanodelivery systems in this book using molecular dynamics (MD) simulations instead. MD simulation is a great tool for the studies of molecular and atomic systems. The descriptions of building an MD simulation program will be covered in Chap. 2, and an example of using the MD simulation program to investigate a nanosystem will be shown and discussed in Chap. 3. We will be focusing on the investigations of the CNT channels as nanodelivery systems for copper atoms in Chaps. 4–6, using MD simulation methods as the principal tool in our investigations.

References

1. Y.L. Mao, X.H. Yan, Y. Xiao, J. Xiang, Y.R. Yang, H.L. Yu, The viability of 0.3 nm diameter carbon nanotubes. Nanotechnology **15**, 1000 (2004)
2. Y. Huang, D.G. Vlachos, J.G. Chen, Synthesis of rigid and stable large-inner-diameter multiwalled carbon nanotubes. RSC Advances **2**, 2685–2687 (2012)
3. G. Editorial, Who should be given the credit for the discovery of carbon nanotubes? Carbon **44**, 1621–1623 (2006)
4. S. Iijima, Helical microtubules of graphitic carbon. Nature **354**, 56–58 (1991)
5. N. Sinha, J.T.-W. Yeow, Carbon nanotubes for biomedical applications. IEEE Trans. Nanobiosci. **4**, 180–195 (2005)
6. Z. Liu, M. Winters, M. Holodniy, H. Dai, siRNA delivery into Human T Cells and Primary Cells with carbon-nanotube transporters. Angew. Chem. Int. Ed. **46**, 2023–2327 (2007)

7. A. Star, E. Tu, J. Niemann, J.-C.P. Gabriel, C.S. Joiner, C. Valcke, Label-free detection of DNA hybridization using carbon nanotube network field-effect transistors. Biophysics **103**, 921–926 (2006)
8. B. Gigliotti, B. Sakizzie, D.S. Bethune, R.M. Shelby, J.N. Cha, Sequence-Independent helical wrapping of single-walled carbon nanotubes by long genomic DNA. Nano Lett. **6**, 159–164 (2006)
9. N.N. Naguib, Y.M. Mueller, P.M. Bojczuk, M.P. Rossi, P.D. Katsikis, Y. Gogotsi, Effect of carbon nanofibre structure on the binding of antibodies. Nanotechnology **16**, 567–571 (2005)
10. R. Wan, J. Li, H. Lu, H. Fang, Controllable water channel gating of nanometer dimensions. J. Am. Chem. Soc. **127**, 7166–7170 (2005)
11. A.T.C. Johnson, C. Staii, M. Chen, S. Khamis, R. Johnson, M.L. Klein, A. Gelperin, DNA-decorated carbon nanotubes for chemical sensing. Physica Status Solidi (b) **243**, 3252–3256 (2006)
12. I. Monch, A. Leonhardt, A. Meye, S. Hampel, R. Kozhuharova-Koseva, D. Elefant, M.P. Wirth, B. Buchner, Synthesis and characteristics of Fe-filled multi-walled carbon nanotubes for biomedical application. J. Phys: Conf. Ser. **61**, 820–824 (2007)
13. R. Hatakeyama, Y.F. Li, T. Kaneko, *Transport properties of p-n junctions created in single-walled carbon nanotubes by Fe encapsulation*, in Nanotechnology, 2007. IEEE-NANO 2007. 7th IEEE Conference on, 2007, pp. 180–184
14. E. Katz, I. Willner, Biomolecule-functionalized carbon nanotubes, applications in nanobioelectronics. Chem Phys Chem **5**, 1084–1104 (2004)
15. S. Iijima, T. Ichihashi, Single-shell carbon nanotubes of 1-nm diameter. Nature **363**, 603–605 (1993)
16. D.S. Bethune, C.H. Klang, M.S. de Vries, G. Gorman, R. Savoy, J. Vazquez, R. Beyers, Cobalt-catalysed growth of carbon nanotubes with single-atomic-layer walls. Nature **363**, 605–607 (1993)
17. M. Kosaka, T.W. Ebbesen, H. Hiura, K. Tanigaki, Annealing effect on carbon nanotubes. An ESR study. Chem Phys Lett **233**, 47–51 (1995)
18. T.W. Ebbesen, T. Takada, Topological and SP3 defect structures in nanotubes. Carbon **33**, 973–978 (1995)
19. R. Saito, M. Fujita, G. Dresselhaus, M.S. Dresselhaus, Electronic structure of chiral graphene tubules. Appl. Phys. Lett. **60**, 2204–2206 (1992)
20. R. Saito, G. Dresselhaus, M.S. Dresselhaus, Tunneling conductance of connected carbon nanotubes. Phys. Rev. **53**, 2044–2050 (1996)
21. Z. Yao, H.W.C. Postma, L. Balents, C. Dekker, Carbon nanotube intramolecular junctions. Nature **402**, 273–276 (1999)
22. L. Chico, V.H. Crespi, L.X. Benedict, S.G. Louie, M.L. Cohen, Pure carbon nanoscale devices: nanotube heterojunctions. Phys. Rev. Lett. **76**, 971–974 (1996)
23. J.C. Charlier, T.W. Ebbesen, P. Lambin, Structural and electronic properties of pentagon-heptagon pair defects in carbon nanotubes. Phys. Rev. **53**, 11108–11113 (1996)
24. C. Jin, K. Suenaga, S. Iijima, Plumbing carbon nanotubes. Nat. Nanotechnol. **3**, 17–21 (2008)
25. I. Hanasaki, A. Nakatani, Water flow through carbon nanotube junctions as molecular convergent nozzles. Nanotechnology **17**, 2794–2804 (2006)
26. M.C.G. Lim, Z.W. Zhong, Effects of fluid flow on the oligonucleotide folding in single-walled carbon nanotubes. Phys. Rev. **80**, 041915-1-8 (2009)
27. Y.F. Li, R. Hatakeyama, J. Shishido, T. Kato, T. Kaneko, Air-stable p-n junction diodes based on single-walled carbon nanotubes encapsulating Fe nanoparticles. Appl. Phys. Lett. **90**, 173127-1-3 (2007)
28. U. Weissker, S. Hampel, A. Leonhardt, B. Büchner, Carbon nanotubes filled with ferromagnetic materials. Materials **3**, 4387–4427 (2010)
29. R.D.R. Meyer, J. Sloan, R.E. Dunin-Borkowski, A.I. Kirkland, M.C. Novotny, S.R. Bailey, J.L. Hutchison, M.L.H. Green, Discrete atom imaging of one-dimensional crystals formed within single-walled carbon nanotubes. Science **289**, 1324–1326 (2000)

30. A. Govindaraj, B.C. Satishkumar, M. Nath, C.N.R. Rao, Metal nanowires and intercalated metal layers in single-walled carbon nanotube bundles. Chem. Mater. **12**, 202–205 (1999)
31. E. Borowiak-Palen, E. Mendoza, A. Bachmatiuk, M.H. Rummeli, T. Gemming, J. Nogues, V. Skumryev, R.J. Kalenczuk, T. Pichler, S.R.P. Silva, Iron filled single-wall carbon nanotubes—a novel ferromagnetic medium. Chem. Phys. Lett. **421**, 129–133 (2006)
32. S. Costa, E. Borowiak-Palen, A. Bachmatiuk, M.H. Rümmeli, T. Gemming, R.J. Kalenczuk, Filling of carbon nanotubes for bio-applications. Physica Status Solidi (b) **244**, 4315–4318 (2007)
33. F.W. Sun, H. Li, K.M. Liew, Compressive mechanical properties of carbon nanotubes encapsulating helical copper nanowires. Carbon **48**, 1586–1591 (2010)
34. W. Han, P. Redlich, F. Ernst, M. Ruhle, Synthesizing boron nitride nanotubes filled with SiC nanowires by using carbon nanotubes as templates. Appl. Phys. Lett. **75**, 1875–1877 (1999)
35. M. Monthioux, Filling single-wall carbon nanotubes. Carbon **40**, 1809–1823 (2002)
36. R. Fan, R. Karnik, M. Yue, D. Li, A. Majumdar, P. Yang, DNA translocation in inorganic nanotubes. Nano Lett. **5**, 1633–1637 (2005)
37. F. Banhart, N. Grobert, M. Terrones, J.-C. Charlier, P.M. Ajayan, Metal atoms in carbon nanotubes and related nanoparticles. Int. J. Mod. Phys. **15**, 4037–4069 (2001)
38. H. Kataura, Y. Maniwa, T. Kodama, K. Kikuchi, K. Hirahara, K. Suenaga, S. Iijima, S. Suzuki, Y. Achiba, W. Krätschmer, High-yield fullerene encapsulation in single-wall carbon nanotubes. Synth. Met. **121**, 1195–1196 (2001)
39. J. Wu, M.-L. Wang, R. Lu, W. Duan, The study on the filling of atoms in a carbon nanotube. Int. J. Mod. Phys. **12**, 1601–1606 (1998)
40. G.Y. Zhang, E.G. Wang, Cu-filled carbon nanotubes by simultaneous plasma-assisted copper incorporation. Appl. Phys. Lett. **82**, 1926–1928 (2003)
41. D. Schebarchov, S.C. Hendy, Capillary absorption of metal nanodroplets by single-wall carbon nanotubes. Nano Lett. **9**, 3668 (2009)
42. D. Schebarchov, S.C. Hendy, Capillary absorption of metal nanodroplets by single-wall carbon nanotubes. Nano Lett. **8**, 2253–2257 (2008)
43. K. Svensson, H. Olin, E. Olsson, Nanopipettes for metal transport. Phys. Rev. Lett. **93**, 145901-1-4 (2004)
44. J.P. Dekker, A. Lodder, J. van Ek, Theory for the electromigration wind force in dilute alloys. Phys. Rev. **56**, 12167–12177 (1997)
45. S. Fujisawa, T. Kikkawa, T. Kizuka, Direct observation of electromigration and induced stress in Cu Nanowire. Jpn. J. Appl. Phys. **42**, L1433–L1435 (2003)
46. F.G. Sen, M.K. Aydinol, Atomistic simulation of self-diffusion in Al and Al alloys under electromigration conditions. J. Appl. Phys. **104**, 073510–073514 (2008)
47. J.W. Kang, H.J. Hwang, Model schematics of a nanoelectronic device based on multi-endo-fullerenes electromigration. Physica E **27**, 245–252 (2005)
48. B.C. Regan, S. Aloni, R.O. Ritchie, U. Dahmen, A. Zettl, Carbon nanotubes as nanoscale mass conveyors. Nature **428**, 924–927 (2004)
49. L.X. Dong, X.Y. Tao, L. Zhang, X.B. Zhang, B.J. Nelson, Nanorobotic spot welding: controlled metal deposition with attogram precision from copper-filled carbon nanotubes. Nano Lett. **7**, 58–63 (2007)
50. D. Ugarte, A. Chatelain, W.A. de Heer, Nanocapillarity and chemistry in carbon nanotubes. Science **274**, 1897–1899 (1996)

Chapter 2
Building an MD Simulation Program

2.1 Introduction

In this chapter, we will focus on how to build a molecular dynamics (MD) simulation program using three important parameters for an atom: forces, velocities and positions. There are many different equations available for calculating the parameters of an atom in an MD simulation program, depending on the level of the accuracy that we would like to achieve and the type of the molecular system that we are studying. Besides that, there are also different methods developed over the years to improve the accuracy of the integrated parameters. Our main focus in this book is to demonstrate the construction of an MD simulation program, using simple yet realistic equations to describe molecular phenomena as accurate as possible. In order to maintain the relevancy of the equations used for the MD simulation program and the system that we are going to discuss in the following chapters, the following sections highlight some basic theories and equations required for constructing an MD simulation program for the analysis of transport phenomena in carbon nanotube channels.

2.2 Ensembles

An MD simulation is a useful tool for predicting the atomic positions, velocities and forces of atoms. The calculations based on statistical mechanics are essential for converting the microscopic information to macroscopic phenomena, such as pressure, heat capacities, energy and other properties. The properties of the systems in MD simulations are calculated in terms of the function of time. By integrating Newton's second law of motion using different integration algorithms, we can determine the atomic trajectories of atoms in space and time using the calculated acceleration values of the atoms.

There are four types of ensembles in MD simulations, which describe the conditions of a particular thermodynamics state of the atomic system [1].

M. C. G. Lim and Z. W. Zhong, *Carbon Nanotubes as Nanodelivery Systems*,
SpringerBriefs in Applied Sciences and Technology,
DOI: 10.1007/978-981-4451-39-0_2, © The Author(s) 2013

- Microcanonical ensemble: This ensemble is characterised by a fixed number of atoms, N, a fixed volume, V and a fixed energy, E. It is also commonly referred to as an NVE ensemble. This ensemble corresponds to an isolated system.
- Canonical ensemble: This is a thermodynamic state, which is characterised by a fixed number of atoms, N, a fixed volume, V and a fixed temperature, T. It is an ensemble commonly referred to as an NVT ensemble.
- Isobaric–isothermal ensemble: The thermodynamic state for this ensemble is characterised by a fixed number of atoms, N, a fixed pressure, P and a fixed temperature, T. It is referred to as an NPT ensemble.
- Grand canonical ensemble: This ensemble comprises a thermodynamic state, which is characterised by a fixed chemical potential, μ, a fixed volume, V and a fixed temperature, T. It is referred to as a μVT ensemble.

2.3 Periodic Boundary Conditions

A periodic boundary condition is an important condition for simulating a small part (computational box) of a large system. The original computational box is defined as the primary cell. The atoms in the primary cells are replicated in the imaginary cells, surrounding all around the primary cell. There are a total of 26 imaginary cells surrounding the primary cell. As atoms move out from one side of the primary cell into the surrounding image cells, the position of the atoms is recalculated and then translated back into the primary cell by allowing the atoms to re-enter to the opposite side of the primary cell. It should be noted that the computational box should be larger than twice the cutoff distance of the interaction potential used in the system.

2.4 Neighbour List

The neighbour list is created in MD simulations, such that the computation speed can be improved by reducing the time required to evaluate the pair interactions in the system. In order to create the list, a loop over particles from $i = 1$ to $i = N - 1$, where N is the number of particles, is examined. The first loop encloses a second loop, in which the second loop covers all the possible neighbours of that particle, from $j = i + 1$ to $j = N$. There will be a total of $N(N - 1)/2$ pairs of atoms, with the consideration of the boundary conditions of the system during the generation of the list. If the distance between two atoms is greater than the cutoff distance, the interactive force is assumed to be zero, and the loop for the jth atom is thus terminated.

2.5 Potential Energy and Forces

2.5.1 Embedded Atom Method

There are various models in the field of MD simulations, which can be used to approximate the energy between two atoms. The interaction between two atoms may be weak or strong, depending on the type of the phase and the element of the atoms being modelled in the simulations. Simple pair potential is the most basic potential that is often used to describe organic molecules, liquid and gaseous atoms. However, for metal atoms, a stronger interaction is required instead to describe the potential energy of the atoms more realistically, and therefore the embedded atom method (EAM) is being considered in the examples in this book for the interaction between metal atoms.

EAM is a method, which describes the potential of metal atoms with the combination of pair potential and electron density embedded functions in its energy equation. One of the differences between the EAM and pair potentials is that pair potentials do not properly account for local density variations. The pair potentials are also not able to consistently describe the forces in particular atomic configurations for metals, even though the total energy is calculated based on the sum of pair potentials between two atoms [2]. In the case of metal atoms due to the EAM, the dominant energy of the atom is calculated based on the energy required to embed an atom into the local electron density caused by the surrounding atoms of the system, which results in many-body effects [3, 4]. The EAM thus consists of the computational simplicity required for defects and amorphous systems, and avoids the ambiguities of the pair potential schemes [5].

The total energy of the EAM [6], E_{tot}, combines the energy, ξ, required for embedding atoms into the homogenous electron gas, and the energy, $\psi(r_{ij})$, for the two-body interaction:

$$E_{\text{tot}} = \sum_{i}^{N} \xi_i(\varsigma_i) + \frac{1}{2} \sum_{i,j}^{N} \psi(r_{ij}). \tag{2.1}$$

r_{ij} is the distance between atoms i and j. ς_i denotes the host electron density at atom i due to the surrounding atoms. $\xi_i(\varsigma_i)$ is the embedded function in the universal form [7] and $\psi(r_{ij})$ is the two-body potential function [8].

2.5.2 Forces on Atoms Due to the EAM Potential

The potential energy of an atom, such as the EAM potential, is differentiated during the simulation in order to determine the force on an atom due to the system. As the positions of the neighbouring atoms change, the overall effect of the homogeneous electron gas caused by the surrounding atoms and pair potentials would influence the force on the atom. Based on the magnitude of the force on the atom, the acceleration of the atom is then determined. The force on the atom is expressed as the differentiation of the EAM potential [9, 10]:

$$\gamma_i = -\sum_{j=1j\neq i}^{N}\left[\frac{d\psi(r_{ij})}{dr_{ij}}+\left(\frac{d\xi_i}{d\left(\sum_j\varsigma(r_{ij})\right)_i}+\frac{d\xi_j}{d\left(\sum_i\varsigma(r_{ji})\right)_j}\right)\left(\frac{d\varsigma(r_{ij})}{dr_{ij}}\right)\right]\frac{\vec{r}_{ij}}{|r_{ij}|},$$

(2.2)

where γ_i represents the force on the ith atom, N is the total number of atoms in the system and j is the number of neighbouring atoms surrounding the ith atom. r_{ij} is the distance between the ith and jth atoms. $\frac{d\psi(r_{ij})}{dr_{ij}}$ is the derivative of the pair potential between the ith and jth atoms with respect to the separation between them. $\frac{d\xi_i}{d\varsigma(r_{ij})_i}$ and $\frac{d\xi_j}{d\varsigma(r_{ij})_j}$ are the derivatives of the 'embedding function' for the ith and jth atoms, respectively. $\varsigma(r_{ij})$ is the electron density provided to the embedded ith atom by its surrounding jth atoms.

The host electron density is first calculated and stored in an array of matrix for all atoms in the simulation, which uses the EAM potential. The derivatives of the 'embedded function' for each of the ith and jth atoms are then computed simultaneously during the simulation, using the stored host electron density values [11]. The algorithm for computing the energy and force of an atom using the EAM potential is elaborated further in Sect. 2.5.5.

2.5.3 Lennard-Jones Potential

The Lennard-Jones (LJ) potential is a pair potential for computing the interactive forces between atoms. One of the applications of this potential in this book is to describe the interaction between metal atoms and the CNT carbon atoms. The expression of the potential energy between metal and carbon atoms is shown as [12]:

$$E_{\text{Cu-C}}(r) = \sum_{i<j}\left[\frac{C_{12}(i,j)}{r_{ij}^{12}} - \frac{C_6(i,j)}{r_{ij}^6}\right],$$

(2.3)

where r is the distance between two atom pairs, and C_6 and C_{12} are constants.

2.5.4 Potential and Force Algorithms

Figure 2.1 shows the algorithm of computing the potential energies and forces for metal atoms in MD simulations using the EAM potential. Before initializing the program, the parameters of the EAM potential are first defined by the users. Next, the atom pairs are chosen according to the neighbour list and the cutoff distance that is defined by the users. The neighbour list will help to ease the computational effort in determining the neighbouring contribution of the electron density and the pair potential for the embedded ith atom in the system. The positions of the atoms are checked against the periodic boundary condition of the system before the calculation of the host electron density is carried out.

The computation of the forces between two atoms is determined based on the cutoff distance between the ith and the jth atoms. This step is essential for increasing the computational speed. According to Eq. (2.2), the force on an atom is dependent on the derivatives of the pair potential and the embedded function of the EAM potential. The host electron density is calculated in the first loop of the algorithm for all the atoms, as shown in Fig. 2.1, because the host electron density for both the ith and the jth atoms is required for the calculation of the derivatives of the embedded function.

Next, the atom pairs are chosen again according to the neighbour list and the cutoff distance, before the computation of the pair potential, remaining embedded function terms, energy and force for each atom is carried out. The calculation is done for each of the atom pairs simultaneously in the second loop of the algorithm. Once the calculations are completed, the values are converted according to the units that have been determined by the user for the system.

2.5.5 Electromigration Forces on Metal Atoms

Electromigration is one of the transport methods for the encapsulated materials in CNTs. The electromigration consists of the direct force and the wind force. The expression of the electromigration force is shown as [13]:

$$\gamma_{\mathrm{em}} = (F_d + F_w)e\beta_f = \left(F_d + \frac{k_m}{R(T)}\right)e\beta_f, \qquad (2.4)$$

where F_d and F_w are the valence equivalents of the direct and wind forces respectively, e is the electron charge, and β_f is the electric field. The electric field is defined as the region, in which a point charge experiences a force exerted by another charge particle. The unit of the electric field is voltage per meter, V/m (Vm^{-1}). T is the temperature in Kelvin and $R(T)$ is the resistivity of the material. The resistivity is temperature dependent and can be described as $R(T) = 2.96 + 0.00768(T - 473)$ $\mu\Omega$cm for copper [14]. The temperature-independent quantity, k_m, for copper is given as $k_m = -33.3$ $\mu\Omega$cm [14, 15].

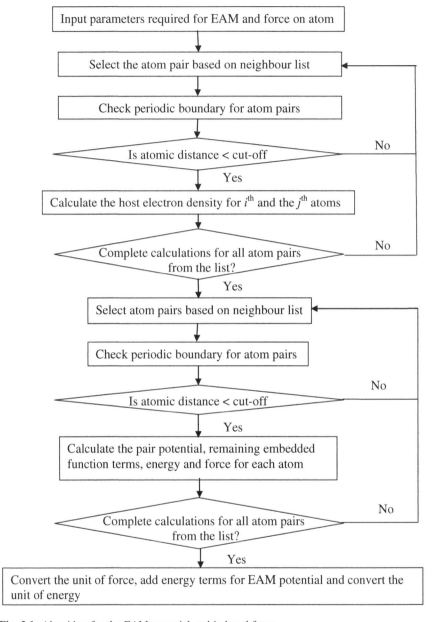

Fig. 2.1 Algorithm for the EAM potential and induced force

Electromigration is a forced atomic diffusion due to an electric field and associated electric current in the metals [16]. In this book, we consider the wind force as the main contributor, which is used as the good measure for the electromigration force, because we assume the contribution of the direct force to be small [14]. On the other hand, we assume that the cluster of metal atoms that we are going to study is a result of the separation from the bulk material due to electromigration, and hence the cluster of metal atoms would not only be influenced by the scattering of electrons, but also be affected by the bulk resistivity of the bulk material.

2.6 The Velocity Verlet Method

In MD simulations, there are several numerical algorithms, which can be used to integrate the equation of motion. The method shown here is known as the velocity Verlet method. The equations for determining the position, velocity and acceleration of an atom are as follows:

$$v\left(t + \frac{\delta t}{2}\right) = v(t) + a(t)\frac{\delta t}{2}, \tag{2.5}$$

$$s_p(t + \delta t) = s_p(t) + v\left(t + \frac{\delta t}{2}\right)\delta t, \tag{2.6}$$

$$a(t + \delta t) = \ddot{s}_p(t + \delta t) = \gamma(s_p(t + \delta t))/m, \tag{2.7}$$

$$v(t + \delta t) = v\left(t + \frac{\delta t}{2}\right) + \frac{1}{2}(a(t + \delta t))\delta t, \tag{2.8}$$

where v is the velocity, a is the acceleration, s_p is the position and t is time. The velocity Verlet method consists of a velocity correction factor at half a time step, which allows the acceleration of the atom over the time step to be taken into consideration, such that the final velocity is better predicted.

The advantage of using the velocity Verlet algorithm is that the velocity term appears directly in the equation of motion to be iterated, so that no extra effort or storage is required. Apart from that, the equations retain the superior numerical precision of the summed form [17]. The velocity Verlet method is the preferred choice for the MD simulations in this book, because it provides convenience for nonequilibrium MD calculations, which need continuous or periodic rescaling of the velocities for atoms.

The velocity of the encapsulated atoms in a CNT channel is computed using Newton's equation of motion. If we assume that each of the atoms is being pushed by an electromigration force, γ_{em}, the resulting equation of motion for the ith atom along the z direction would be:

$$m\frac{\mathrm{d}}{\mathrm{d}t}\left(v_i - u_i\right)_z = \gamma_{\mathrm{em}} + \sum_{j=1, j\neq i}^{N} \left(\nabla E_{ij}\right)_z, \tag{2.9}$$

where N is the number of atoms translocating along the CNT, m is the mass of an atom, v is the velocity of the atom at time $t + \mathrm{d}t$, u is the velocity of the atom at time t and E is the potential energy between two atoms.

In order to maintain a constant temperature in the system during simulations, the velocity of the moving atoms is required to undergo a rescaling process which takes place before the velocity Verlet method is implemented. According to Eq. (2.9), the velocity of the moving atoms consists of forces due to the potential energy and the electromigration. If we were to rescale the velocity term based on this relationship, the drift velocity of the moving atoms would be affected significantly. In view of this problem, a new set of velocities is calculated for the purpose of computing the rescaling values for adjusting the temperature of the system. The additional set of velocities is computed as:

$$w_{z,i} = w_{z,i}^{\mathrm{old}} + \frac{\Delta t}{m}\left\{\sum_{j=1, j\neq i}^{N}\left(\nabla E_{ij}\right)_z\right\}, \tag{2.10}$$

where N is the number of atoms translocating along the CNT, m is the mass of an atom, w is the velocity of the atom at time $t + \mathrm{d}t$, w^{old} is the velocity of the atom at time t and E is the potential energy between two atoms. The velocity at time $t + \mathrm{d}t$ is equal to the velocity of the moving atom at time t plus the forces due to the potential energy between atoms. This additional velocity term would ensure that the rescaling process would only adjust the thermal contribution to the atoms instead.

2.7 Simulation Algorithm

Figure 2.2 shows the algorithm of an MD simulation. The numbers indicated at the sections of the algorithm represent the steps at which the processes would take place during the simulation. Step 1 defines the variables and constants for the equations and conditions necessary for performing the MD simulations. Step 2 initiates the coordinates, velocities and accelerations of the atoms in the system. The initial neighbour list for atom pairs and the host electron density at equilibrium is created at this stage. The simulation loop for the numerical calculations of MD starts at step 3. Within this loop, the velocity Verlet method, the velocity rescaling method for controlling temperature, periodic boundary conditions, the EAM potential model, the LJ potential and the forces on atoms are all executed for every time steps and every atom.

At step 5, the positions of the atoms are first updated using the velocity Verlet algorithm. Next, the new positions of the atoms are adjusted based on periodic

Fig. 2.2 Simulation
algorithm

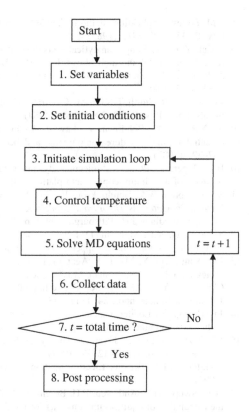

boundary conditions. The neighbour list, the EAM potential and the Verlet algorithm are updated accordingly, such that the velocity of the translocated atoms can be re-established and the accelerations due to the resultant forces can be derived. The information is collected in step 6 for the purpose of executing additional calculations for determining other properties and functions such as the mean square displacement, pair correlation function, layer structure factor, mean square vibrational amplitude and to display values such as energies and temperatures of the system for references. The final consolidation of the data and visual display of the system are carried out after the completion of the simulation at step 8.

References

1. G.S. Rushbrooke, *Introduction to Statistical Mechanics* (Clarendon Press, Oxford, 1960)
2. D.Y. Lo, T.A. Tombrello, M.H. Shapiro, Theoretical studies of ion bombardment: many-body interactions. J. Vac. Sci. Technol., A **6**, 708–711 (1988)
3. M.S. Daw, M.I. Baskes, Semiempirical quantum mechanical calculation of hydrogen embrittlement in metals. Phys. Rev. Lett. **50**, 1285–1288 (1983)
4. M.S. Daw, M.I. Baskes, Embedded-atom method: derivation and application to impurities, surfaces, and other defects in metals. Phys. Rev. B. **29**, 6443–6453 (1984)

5. S.M. Foiles, Application of the embedded-atom method to liquid transition metals. Phys. Rev. B. **32**, 3409–3415 (1985)

6. J. Cai, Y.Y. Ye, Simple analytical embedded-atom-potential model including a long-range force for fcc metal and their alloys. Phys. Rev. B. **54**, 8398–8410 (1996)

7. A. Banerjea, J.R. Smith, Origins of the universal binding-energy relation. Phys. Rev. B. **37**, 6632–6645 (1988)

8. J.H. Rose, J.R. Smith, F. Guinea, J. Ferrante, Universal features of the equation of state of metals. Phys. Rev. B. **29**, 2963–2969 (1984)

9. K. Narayan, K. Behdinan, Z. Fawaz, An engineering-oriented embedded-atom-method potential fitting procedure for pure fcc and bcc metals. J. Mater. Process. Technol. **182**, 387–397 (2007)

10. H.E. Alper, P. Politzer, Molecular dynamics simulations of the temperature-dependent behavior of aluminum, copper, and platinum. Int. J. Quantum Chem. **76**, 670–676 (2000)

11. S.J. Plimpton, B.A. Hendrickson, Parallel Molecular Dynamics with the Embedded Atom Method, in *Materials Theory and Modelling*, ed. by J. Broughton, P. Bristowe, J. Newsam (MRS Proceedings 291, Pittsburgh, 1993), p. 37

12. S. Dorfman, K.C. Mundim, D. Fuks, A. Berner, D.E. Ellis, J. Van Humbeeck, Atomistic study of interaction zone at copper-carbon interfaces. Mater. Sci. Eng., C **15**, 191–193 (2001)

13. J.M. Seminario, Y. Ma, L.A. Agapito, L. Yan, R.A. Araujo, S. Bingi, N.S. Vadlamani, K. Chagarlamudi, T.S. Sudarshan, M.L. Myrick, P.E. Colavita, P.D. Franzon, D.P. Nackashi, L. Cheng, Y. Yao, J.M. Tourc, Clustering effects on discontinuous gold film nanocells. J. Nanosci. Nanotechnol. **24**, 1–11 (2004)

14. J.P. Dekker, A. Lodder, Calculated electromigration wind force in face-centered-cubic and body-centered-cubic metals. J. Appl. Phys. **84**, 1958–1962 (1998)

15. J.P. Dekker, A. Lodder, J. van Ek, Theory for the electromigration wind force in dilute alloys. Phys. Rev. B. **56**, 12167–12177 (1997)

16. D.G. Pierce, P.G. Brusius, Electromigration: a review. Microelectron. Reliab. **37**, 1053–1072 (1997)

17. W.C. Swope, H.C. Andersen, P.H. Berens, K.R. Wilson, A computer simulation method for the calculation of equilibrium constants for the formation of physical clusters of molecules: application to small water clusters. J. Chem. Phys. **76**, 637–649 (1982)

Chapter 3
Sample of an Application of an MD Simulation Program

3.1 Introduction

The MD simulation program is a useful tool for carrying out analyses on a molecular system, in which the progressive changes of the molecular positioning and energy are described over the progression of time steps. Chapter 3 shows and explains how the MD simulation program can be applied and used to carry out the analyses of a molecular/atomic system. Results from the simulations are also presented to give a realistic description of how the system can be analysed using MD methods. The molecular system created for the discussion of this chapter is made up of Al atoms.

The molecular system in this chapter is modelled as a simple rectangular Al block, with the top and bottom surfaces of the block being exposed to air, and the sides of the block connected to a larger Al system. The same Al block is also modelled as a small part of a larger system by surrounding this small part with more Al atoms using the periodic boundary condition method described in Chap. 2. We will be using this simple model to investigate the surface premelting of an Al system, where the exposed area (surface) consists of Al atoms arranged in the (110) configuration. The computational methods section will describe the information required for carrying out MD calculations.

3.2 Simulation Details for a Bulk Al Slab and an Al(110) Surface

MD simulations are performed in this section to study an Al(110) surface. The system consists of an Al fcc crystal with two free surfaces on the (110) planes. The slab consists of 432 atoms, which are divided into 12 layers, 36 atoms each. The dimensions of the slab are $22.28 \times 15.75 \times 15.75$ Å^3 in the x, y and z directions respectively. There are two analyses carried out in this section: the bulk analysis for the Al system and the surface analysis for the Al(110) surface.

M. C. G. Lim and Z. W. Zhong, *Carbon Nanotubes as Nanodelivery Systems*,
SpringerBriefs in Applied Sciences and Technology,
DOI: 10.1007/978-981-4451-39-0_3, © The Author(s) 2013

Periodic boundary conditions are applied in the x, y and z directions for the bulk analysis. On the other hand, periodic boundary conditions for the surface analysis are imposed only in the directions parallel to the surface (x and y directions) of the slab so that diffusion of atoms along the z direction at high temperatures is possible.

The atoms in the system are modelled using the EAM potential with a cutoff distance of $r_{cut} = 1.65a_0$ ($a_0 = 4.05$ Å) [1]. The temperature of the Al slab is controlled during simulations. The temperature of the system is maintained using the velocity rescaling [2]:

$$\frac{v^n}{v^o} = \sqrt{\frac{T_{tr}}{T}}, \tag{3.1}$$

where v^n is the velocity of the atoms after rescaling, v^o is the velocity of the atoms before rescaling, T_{tr} is the targeted temperature and T is the actual temperature of the system.

The temperature of the system is generated based on the average kinetic energy of the atoms [3]:

$$\sum_{i=1}^{N} \frac{1}{2} m_i \sum_{\alpha=x,y,z} \bar{v}_{\alpha,i}^2 = \frac{3}{2} N k_b T. \tag{3.2}$$

N is the number of the atoms in the system and k_b is the Boltzmann constant. The neighbour list is used in this simulation to improve the efficiency of the program.

The simulation cases for the Al system are divided into bulk and surface analyses. There are 9 simulation cases in this section. The temperatures assigned for the Al bulk analysis are 240, 584 and 708 K, whereas the temperatures assigned for the Al(110) surface analysis are 240, 400, 500, 579, 710 and 810 K. The total time simulated for each case is 12 ps. The time step for each simulation is set at 2 fs.

3.3 The Atomic Behavior of a Bulk Al Slab and an Al(110) Surface

The structure of the Al slab and the dynamics of the Al atoms on the surface are verified using the pair correlation function, and the mean square displacement of the atoms in the z direction, perpendicular to the slab surface. The pair correlation function (the probability distribution for the distance between two atoms) for the Al(110) surface is compared with the pair correlation function for the bulk Al.

Figure 3.1 shows the surface dynamic of the Al slab at different temperatures. The black atoms represent the positions of atoms after the simulation is performed at a specified temperature. The surfaces of the slab are located at the left and right sides of the slab, perpendicular to the z direction. The Al atoms are in the perfect

Fig. 3.1 Dynamics of Al
atoms on the surface of the
slab at various temperatures.
The *black* atoms represent the
positions of the Al atoms at
the specified temperature

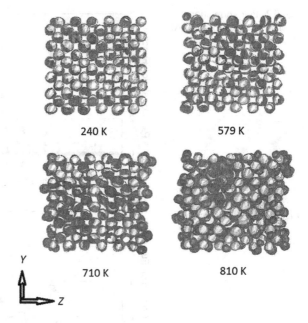

240 K   579 K

710 K   810 K

lattice position at temperatures up to 579 K, as shown in Fig. 3.1. As the temperature increases, the Al atoms become more disordered, especially at 710 K.
This phenomenon occurs at a lower temperature than the expected premelting
temperature of 810 K.

In order to confirm this preliminary finding, pair correlation functions are
plotted for both the bulk and surface Al atoms. The pair correlation results for the
bulk Al at 240 K, the Al(110) surface at 240 K and the bulk Al at 584 K are shown
in Figs. 3.2, 3.3 and 3.4 respectively. Figures 3.5, 3.6 and 3.7 show the results of
the pair correlation function at the Al(110) surface at 579 K, the bulk Al at 708 K
and the Al(110) surface at 710 K, respectively.

The periodic structure of the pair correlation function described at 708 K in
Fig. 3.6, for the bulk Al is quite reasonable, compared to Schommers's finding [4].
Figure 3.7 shows the pair correlation function analysis for the Al(110) surface at
710 K, and the difference between the highest and lowest peaks for the pair
correlation function is about 1.5. Based on Fig. 3.7, the anharmonicity in the figure
suggest that the premelting phenomenon might occur near 710 K.

The layer structure factors, $s_x(K)$, for the outermost four layers of the Al(110)
surface are presented in Fig. 3.8. The layer structure factor is expressed as [5]:

$$s_{l,x}(K) = \left\langle \frac{1}{N_l} \sum_{i \in l} e^{iKr_{v,i}} \right\rangle, \tag{3.3}$$

Fig. 3.2 Pair correlation
function analysis for bulk Al
at 240 K

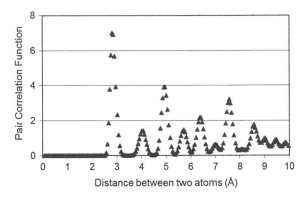

Fig. 3.3 Pair correlation
function analysis for the
Al(110) surface at 240 K

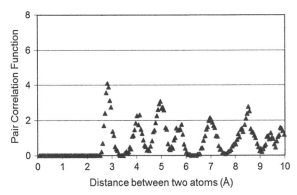

Fig. 3.4 Pair correlation
function analysis for bulk Al
at 584 K

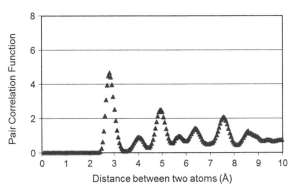

where $K = (2\pi/a_0)(0,0,2)$ is the reciprocal lattice vibration vector along the rows
in the [001] direction, $r_{v,i}$ is the radius vector of the ith atom, a_0 is the lattice
parameter and N_l is the number of atoms in the lth atomic layer. The angular
bracket represents an averaging over the simulated trajectories. For a perfectly
ordered fcc structure at 0 K, $s_x(K)$ is equal to 1. On the other hand, $s_x(K)$ is equal to
0 for a disordered structure such as liquid. According to Fig. 3.8, the layer
structure factors for the four outermost layers decrease with temperature almost

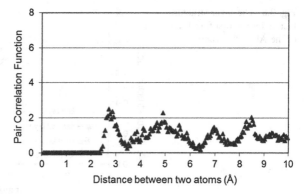

Fig. 3.5 Pair correlation function analysis for the Al(110) surface at 579 K

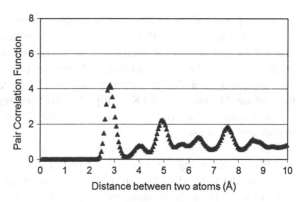

Fig. 3.6 Pair correlation function analysis for bulk Al at 708 K

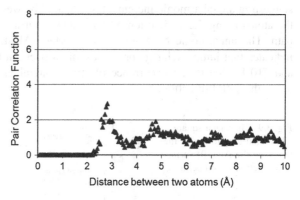

Fig. 3.7 Pair correlation function analysis for the Al(110) surface at 710 K

linearly until around 710 K. The vibrational amplitudes of the Al atoms at the four outermost layers increase at around 710 K, which suggests that the premelting phenomenon of the first four layers becomes more significant at temperatures higher than 710 K.

The gradual increase in the slope of the graph for the layer structure factors at temperatures between 600 and 710 K, suggests that a loss of order in the atomic arrangement for the outermost four layers had occurred, which is associated with

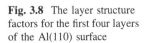

Fig. 3.8 The layer structure
factors for the first four layers
of the Al(110) surface

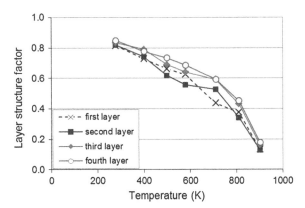

the melting phenomenon. The first four layers of the Al slab melted at a temperature approximately <100 K below the bulk melting temperature of 800 K. When there is a significant decrease in the layer structure factor, it indicates that the surface atoms begin to shift from their regular positions and become adatoms. This process leads to the formation of vacancies. The occurrence of the disorder of the surface Al atoms at 710 K in Fig. 3.8 therefore represents the complex diffusion events on the Al(110) surface, that leads to the melting phenomenon as temperature increases.

Figure 3.9 shows the mean square vibrational amplitudes of the atoms on the Al(110) surface along the x, y and z directions, which correspond to the [001], [1$\bar{1}$0] and the [110] directions respectively. The mean square vibrational amplitudes of the surface atoms along the x and y directions increase greatly at 710 K, suggesting an anharmonic increment. Compared between the x, y and z directions, the atoms along the y direction on the Al(110) surface have the easiest diffusion path. The anharmonic behaviour of the surface Al atoms along the x direction indicates that jump-exchange processes might occur at temperatures between 650 and 710 K, due to the occurrence of strong lateral vibrations of the row atoms along the [001] direction.

Fig. 3.9 The mean square
vibrational amplitudes of the
atoms on the Al(110) surface
as a function of temperature

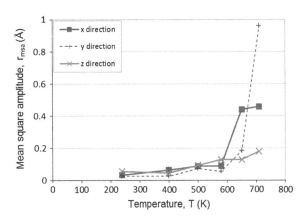

The vibrational amplitudes of the atoms perpendicular (z direction) to the Al(110) surface are within the range of the nearest-neighbour distance ($=2^{-1/2}a$, where a is the lattice constant), as shown in Fig. 3.9. This finding suggests that the EAM potential resulted in ordered oscillation of the surface atoms, perpendicular to the Al(110) surface at temperatures lower than the bulk melting temperature.

The layer diffusion coefficient of the surface atoms which are parallel to the Al(110) surface is expressed as [6],

$$\lim_{t\to\infty}\langle r^2(t)\rangle = 4d_c t, \tag{3.4}$$

where

$$r^2(t) = \left\langle \frac{1}{N_k}\sum_{i\in k} r_i(t+\Delta t) - r_i(t) \right\rangle, \tag{3.5}$$

in which N_l is the number of atoms on the surface layer, d_c is the diffusion coefficient, the angular bracket, $\langle\,\rangle$ represents the averaging over time origins (Δt), and $r^2(t)$ is the mean square displacement. The difference between the diffusion rate and the mean square vibrational amplitude is that the diffusion rate of the surface atoms is influenced by the diffusion coefficient, whereas the mean square vibrational amplitudes vary based on the temperature of the system. By using the diffusion coefficient in Eq. (3.4), the diffusion constant for the outermost layer at 810 K is found to be about 11.76×10^{-5} cm^2/sec, which is close to the diffusion constant of 9.6×10^{-5} cm^2/sec at 810 K as reported by Schommers [4]. The surface premelting of Al (110) has hence occurred at a temperature lower than 810 K, based on the potential used in this example.

The equilibrium melting temperature of the Al material (which is modelled using the EAM) is investigated using the Lindemann index. The Lindemann index is often used to determine the melting temperatures of bulk solids. The Lindemann criterion states that in order to consider a bulk matter as melted, the root mean square (rms) bond length for the thermal fluctuation of the solid should increase more than 10–15 % of the original value, or when the Lindemann index value increases suddenly at a particular temperature [7, 8]. The Lindemann index is the

Fig. 3.10 Lindemann index for the Al material modelled by the EAM potential

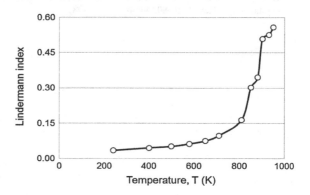

quotient of the rms distance of an atom from its equilibrium position, and the averaged nearest neighbor distance of that particular atom [9]. The equilibrium melting temperature for the Al material modelled in this chapter is around 810 K, as shown in Fig. 3.10. The degree of the undercooling affects the premelted layer width of a material, as suggested by Song et al. [10]. As the undercooling decreases, the layer width of the premelting surface would increase. Lu and Li have reported that the critical undercooling temperature for Al is about 0.16 % below the melting temperature [11]. Based on the equilibrium melting temperature of 810 K which is deduced from Fig. 3.10, the critical undercooling temperature for the Al material would then be about 680 K.

References

1. J. Cai, Y.Y. Ye, Simple analytical embedded-atom-potential model including a long-range force for fcc metal and their alloys. Phys. Rev. B **54**, 8398–8410 (1996)
2. W.-J. Chang, Molecular-dynamics study of mechanical properties of nanoscale copper with vacancies under static and cyclic loading. Microelectron. Eng. **65**, 239–246 (2003)
3. K. Narayan, K. Behdinan, Z. Fawaz, An engineering-oriented embedded-atom-method potential fitting procedure for pure fcc and bcc metals. J. Mater. Process. Technol. **182**, 387–397 (2007)
4. W. Schommers, Structure and dynamics of the Al(110) surface: a molecular dynamics study. Mod. Phys. Lett. B **10**, 963–972 (1996)
5. D.J. Shu, D.Y. Sun, X.G. Gong, W.M. Lau, A molecular-dynamics study of the anisotropic surface-melting properties of Al(110). Surf. Sci. **441**, 206–212 (1999)
6. R.N. Barnett, U. Landman, Surface premelting of Cu(110). Phys. Rev. B **44**, 3226–3239 (1991)
7. F. Calvo, F. Spiegelmann, Mechanisms of phase transitions in sodium clusters: From molecular to bulk behavior. J. Chem. Phys. **112**, 2888–2908 (2000)
8. Y. Shibuta, T. Suzuki, Melting and nucleation of iron nanoparticles: A molecular dynamics study. Chem. Phys. Lett. **445**, 265–270 (2007)
9. Z.H. Jin, P. Gumbsch, K. Lu, E. Ma, Melting mechanisms at the limit of superheating. Phys. Rev. Lett. **87**, 055703 (2001)
10. H. Song, S.J. Fensin, M. Asta, J.J. Hoyt, *A molecular dynamics simulation of (110) surface premelting in Ni*, Scripta Mater. **63**, 128–131
11. K. Lu, Y. Li, Homogeneous nucleation catastrophe as a kinetic stability limit for superheated crystal. Phys. Rev. Lett. **80**, 4474–4477 (1998)

Chapter 4
Carbon Nanotube Channels as Nanodelivery Systems

4.1 Introduction

The investigation of CNT channels as nanodelivery systems will be carried out in this chapter describing the translocation of copper atoms along these channels. We will be using the MD simulation as a tool to carry out the modelling and analyses for the CNT systems. The simulation conditions in this section consist of the descriptions for the CNT systems, conditions for the simulation and parameters for the additional equations required for the analyses.

The CNT system is made up of a (5,5)–(4,4) CNT channel with copper atoms encapsulated within it. The simulations focus on the translocation of copper atoms due to electromigration. As copper atoms are transported along the channels, the resulting changes to the temperatures, energies and forces of the copper atoms are presented to illustrate the flow behaviour of the copper atoms along CNT channels with a junction feature.

4.2 Transportation of Copper Atoms Along a (5,5)–(4,4) CNT Channel

In this section, the transportation of copper atoms along a CNT channel is performed using MD simulations. In the simulations, the CNT channel is modelled as a nonpolarised, single-walled and armchair CNT channel. The CNT channel consists of one CNT junction, which is formed due to the joining of two nanotubes of different diameters. The connecting joint of the CNT channel is connected by a pentagon and heptagon pair of carbons [1]. For this simulation case study, a (5,5)–(4,4) CNT channel with 596 carbon atoms is constructed. The CNT channel has a diameter of 6.8 Å for the (5,5) nanotube section and a diameter of 5.4 Å for the

M. C. G. Lim and Z. W. Zhong, *Carbon Nanotubes as Nanodelivery Systems*, SpringerBriefs in Applied Sciences and Technology, DOI: 10.1007/978-981-4451-39-0_4, © The Author(s) 2013

(4,4) nanotube section. The (5,5)–(4,4) CNT channel is approximately 82 Å in length and spans between −41 and 41 Å along the z coordinate. The carbon atoms are fixed in space.

There are 20 copper atoms encapsulated in the (5,5)–(4,4) CNT channel. Energy minimisation and equilibrium MD runs for the copper atoms are made, before the nonequilibrium MD simulations are performed. The copper atoms are modelled using the EAM potential [2] with the cutoff distance of $r_{cut} = 1.65a_0$ ($a_0 = 3.615$ Å).

The interaction between copper and carbon, $E_{Cu-C}(r)$, can be expressed using Eq. 2.3, where r is the distance between the copper and carbon atoms, $C_6 = 41.548$ (eVÅ^6) and $C_{12} = 2989.105$ (eVÅ^{12}). The cutoff distance is set at 10 Å [3]. A bias voltage is applied along the CNT channel to drive the copper atoms. The driving forces (γ_{em}) exerted on the copper atoms by the electromigration process can be expressed with Eq. 2.4. Since the wind force can be used as a good measure for the main contribution to the driving force due to electromigration in this example, we therefore consider only the wind force for simplicity in this study.

In the MD simulations, the electromigration force is induced towards the − z direction because copper atoms would move along the opposite direction of the applied electromigration force. The bias voltages are set at 2, 4, 6, 8 and 10 V. The repulsive effect between atoms is represented by the positive energies or forces in the simulations, whereas the attractive effect between atoms are represented by the negative values of the energies and forces.

During the simulations, the temperature of the copper atoms would increase due to the effect of the resistive heating in electromigration. In order to maintain the temperature of the system, the modified velocity rescaling method is used in this section to control the outcome of the temperature after copper atoms are induced with the electromigration forces. The equation for the modified velocity rescaling method is expressed as:

$$\frac{u_i^{new}}{u_i^{old}} = \sqrt{\frac{T_t}{T_{e,i}}}, \tag{4.1}$$

where u_i^{new} is the resulting velocity of the ith atom after rescaling, u_i^{old} is the velocity of the ith atom before rescaling, T_t is the targeted temperature and $T_{e,i}$ is the temperature of the ith moving atom. $T_{e,i}$ is calculated based on Eq. 3.2, whereby the velocity term in Eq. 3.2 is generated using Eq. 2.10. Both the velocity Verlet method and the neighbour list are used in the algorithm of the simulations for the translocation of the copper atoms along the CNT channel.

The simulation cases are setup by assigning the initial temperatures of copper atoms at 373, 673, 873, 1073 and 1273 K, which are below the melting temperature of copper. The time step for the simulations is set at 1 fs, with a total duration of 20 ps for each of the simulation studies. The duration of the simulation is set in such a way that it is sufficient for the copper atoms to translocate along the entire CNT channel.

4.3   Analyses of the Dynamical Behaviour of Copper Atoms in a (5,5)–(4,4) CNT Channel

The flow processes of copper atoms along the (5,5)–(4,4) CNT channel is investigated in this section, focusing on the different flow phenomena due to different temperatures. Copper atoms are initially located at the wider end of the channel. A bias voltage is then applied along the CNT channel to induce the electromigration force for the copper atoms to propel from the wider end to the narrow end of the channel. As the temperature and the magnitude of the bias voltage vary in different case studies, the resulting flow of the copper mass is also affected. Based on the conditions of the simulations in this study, the threshold voltage for the translocation of the copper mass along the (5,5)–(4,4) CNT channel is 0.25 V. Apart from the flow conditions, the other contributing factor for the change in the flow behaviour of copper atoms is the obstruction of the CNT junction to the flow processes.

Figure 4.1 shows the translocation of copper atoms along a (5,5)–(4,4) CNT channel at 873 K, due to a bias voltage of 2 V. The copper mass is transported as a whole along the channel, with no copper atom separated from the group of copper mass. When the copper mass reaches the junction, the copper atoms are "pushed" towards the junction due to the driving force and the distances between copper atoms are shortened, as shown in Fig. 4.1b. The cause of the "squeezing" effect on the copper mass is mainly due to the combination of the electromigration force and the blockage of the CNT junction. As the driving force continues to induce onto the copper atoms, the copper atoms are then realigned and forced out of the junction into the narrower region of the channel, which is shown in Fig. 4.1c.

Figure 4.1d shows the increase in the length of the copper mass in the downstream of the channel due to the decrease in the cross-sectional diameter of the downstream section of the channel. However, it is observed that there is a slight difference between the spacing of the copper atoms at the leading and trailing ends of the copper mass existing at the (4,4) CNT. As shown in Fig. 4.1e, the copper atoms at the leading end have a narrower spacing, compared to the copper atoms which are closer to the CNT channel junction. As the copper mass is obstructed by the channel junction during the translocation process, no copper atoms are detached from the copper mass and move into the (4,4) CNT separately. The reason is the magnitude of the electromigration force is strong enough to push the entire copper mass through the CNT junction by overcoming the energy barrier. The ratio between the upstream and downstream channel diameter is closed to 1, and hence the difference between them is small.

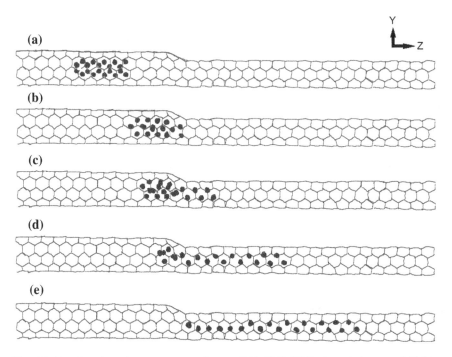

Fig. 4.1 Translocation of copper atoms along a (5,5)–(4,4) CNT channel due to a 2 V bias voltage at 873 K. Figures depicted the snapshots from the simulation at (**a**) 1.6 ps, (**b**) 3.4 ps, (**c**) 4.4 ps, (**d**) 6.7 ps and (**e**) 9.0 ps respectively

The change in the temperature of the copper mass during the translocation processes shows how an electromigration would result in the increase of the temperature for copper atoms, compared to that of the initial temperature assumed for the channel and the copper mass. The increase in the temperature suggests the presence of the effect of Joule heating in the copper mass. Figure 4.2 shows the change in the temperature of copper atoms along the channel during simulations. At 373 K, the increase in the bias voltage causes the temperature of copper atoms to increase the most, compared to the other system with higher initial temperatures. A bias voltage of 10 V increases the initial temperature by almost 800 K in copper atoms, as shown in Fig. 4.2a.

In Fig. 4.2e, it is observed that the bias voltage of 2 V at 1273 K causes the initial temperature to increase by about 25 K, whereas a bias voltage of 10 V results in an increase of about 230 K in the copper atoms. When there is an increase in the bias voltage, the kinetic energy of the copper atoms is also increased, therefore causing the temperature of the copper mass to elevate. On the other hand, when the copper mass is blocked by the channel junction (at about 4 ps as shown in Fig. 4.2), the velocity of the copper atoms is reduced, resulting in a lower kinetic energy, hence effectively lowering the change in temperature. For systems with initial temperatures of 673 K and above, the maximum increase

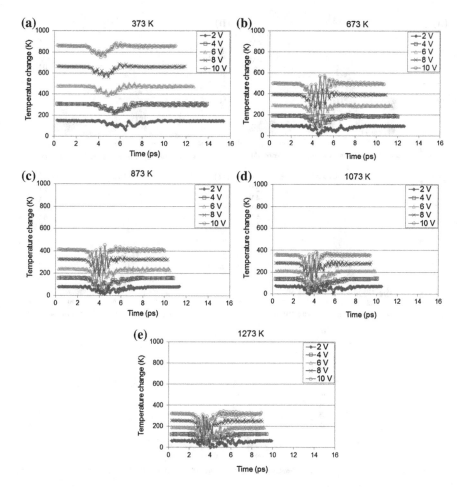

Fig. 4.2 Temperatures of the copper atoms during the translocation processes. The temperature of the CNT channel with copper atoms is initially assumed to be (**a**) 373 K, (**b**) 673 K, (**c**) 873 K, (**d**) 1073 K and (**e**) 1273 K respectively

caused by the bias voltage of 10 V is less than 600 K, whereas a bias voltage of 2 V results in temperature differences of less than 100 K. The highest temperature achieved in the simulations of the flow of copper atoms in the (5,5)–(4,4) CNT channel is ranged between 1200 and 1500 K, for the bias voltage of 10 V, as suggested in Fig. 4.2a.

In order to have a better understanding of the relationship between the copper mass and the CNT channel during the translocation process, the change in the interactive forces between the copper mass and the CNT channel during simulations is investigated. The position of the copper mass along the CNT channel is defined according to the placement of the centre of mass (COM) of the copper atoms along the z-axis of the channel. Figure 4.3 shows the interactive forces

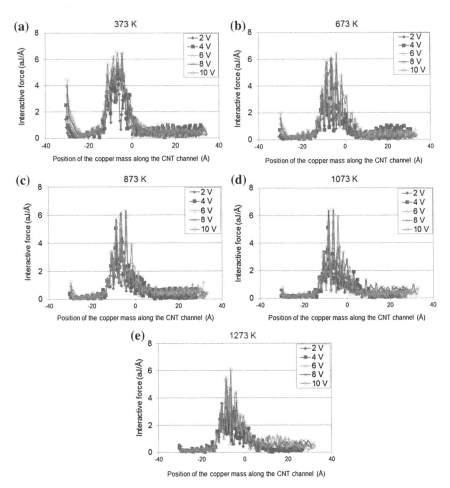

Fig. 4.3 Interactive forces between the copper mass and the CNT channel. The temperature of the CNT channel with copper atoms is initially assumed to be (**a**) 373 K, (**b**) 673 K, (**c**) 873 K, (**d**) 1073 K and (**e**) 1273 K respectively

between the copper mass and the CNT channel. There are three main segments of the CNT channel designated for the analyses of the interactive forces. The three segments are: the (5,5) CNT segment (between −40 and −11 Å), the channel junction (between −11 and 2 Å) and the (4,4) CNT segment (between 2 and 40 Å).

When the copper mass is transported along the (5,5) CNT segment between −30 and −11 Å, the interactive forces between the copper mass and the CNT channel are less than 0.5 aJ/Å. However, the interactive forces between the channel and copper increase as the copper mass approaches the CNT junction region. The increase in the interactive forces at the junction indicates strong repulsive forces being generated between the copper atoms and the CNT channel junction.

As compared between figures (a–e) in Fig. 4.3, it is observed that the highest interactive force of about 6.5 aJ/Å occurs when copper is near the junction, with a surrounding temperature of 373 K and a bias voltage of 10 V, as shown in Fig. 4.3a. At 1273 K, the thermal vibration of the copper atoms is large because the temperature of the copper atoms is close to the melting temperature. When copper atoms are blocked by the CNT junction at this temperature, the reconfiguration of the copper atoms becomes easier and hence lower repulsive forces are generated. On the other hand, the atomic arrangements of copper atoms at lower temperatures are more organised and compact due to smaller thermal vibrations. When the bias voltage is high (high driving force) at a low temperature, the repulsive forces between the CNT channel junction and the copper mass would be higher too. According to Fig. 4.3, the interactive forces between copper atoms and the straight segments of the channel ((4,4) and (5,5) CNT) are quite low, regardless of the temperature or the magnitude of the driving force.

The potential energy of the copper mass is shown in Fig. 4.4, in order to have a better understanding of the atomic arrangement of the copper atoms during simulations. The potential energy of copper atoms indicates the possible changes to the atomic configuration of the copper mass. According to Fig. 4.4, a significant change to the atomic configuration of copper atoms occurs at the CNT junction, which is indicated as a large variation of the potential energy. The potential energy of copper at the (5,5) CNT segment and the (4,4) CNT segment however is almost similar for all the case studies at different temperatures.

Figure 4.4 also shows that the potential energy of the copper mass at the junction is higher when the surrounding temperature is low, and when the magnitude of the driving force is high. At a low temperature, the thermal vibration of copper atoms is small, resulting in lesser atomic movements. When the copper mass moves towards the CNT, the electromigration force and the obstruction of the CNT junction result in an increase in the potential energy of the copper mass. Due to the difficulty of the reorganisation of the copper atoms at lower temperatures, the potential energy becomes higher because the repulsion between copper atoms is stronger when the copper mass is "squeezed" towards the junction.

Based on the observation made in Fig. 4.4, the potential energy between −30 and −12 Å (position of the copper mass) is negative. This suggests that the copper atoms are in a compact state and the attractive energy dominates among them. The increase in the potential energy as copper atoms pass through a CNT junction indicates that the atomic distances between copper atoms are closer than the initial atomic separation of copper atoms at the (5,5) CNT. The narrowing of atomic separation between copper atoms at the CNT junction would result in the repulsive effect among the copper atoms.

Since the (4,4) CNT segment is narrower than the (5,5) CNT segment, the mobility of copper atoms at the (4,4) CNT segment would be more difficult. As a result, the collision between copper atoms would be more frequent and more repulsion occurs between them. The outcome of this phenomenon is reflected as a higher potential energy of copper atoms at the (4,4) CNT segment, compared to the (5,5) CNT segment. The difference in the energies also reflects the changes in

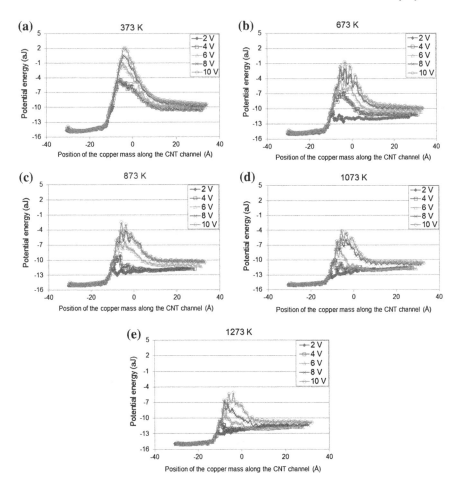

Fig. 4.4 The change in the potential energy for copper atoms along the CNT channel. The temperature of the CNT channel with copper atoms is initially assumed to be (**a**) 373 K, (**b**) 673 K, (**c**) 873 K, (**d**) 1073 K and (**e**) 1273 K respectively

the atomic configuration of the copper mass, and the CNT junction acts as a buffer region for the change to occur.

Figure 4.5 shows the resistive force acting on the copper mass by the CNT channel wall. Among the figures in Fig. 4.5, Fig. 4.5e shows the resistive forces with the least fluctuations. The maximum resistive force acting on the copper mass by the CNT channel junction at 373 K is about 6 aJ/Å, which is larger than the maximum resistive force at 1273 K under the same bias voltage of 10 V. The resistive forces at the (5,5) CNT and (4,4) CNT segments are close to 0 aJ/Å, as shown in Fig. 4.5a–e. Since the resistive forces are low, the difference in the resistive effect between these two segments is quite insignificant.

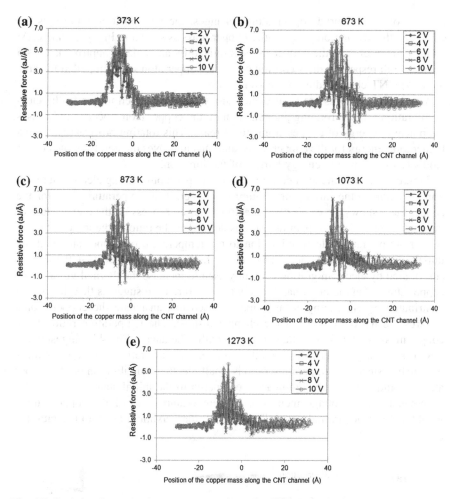

Fig. 4.5 Resistive forces on the copper mass along the CNT channel. The temperature of the CNT channel with copper atoms is initially assumed to be (**a**) 373 K, (**b**) 673 K, (**c**) 873 K, (**d**) 1073 K and (**e**) 1273 K respectively

4.4 Analyses of the Atomic Arrangement of Copper Atoms Under Electromigration Conditions

This section of the book focuses on the continuation of the investigation for the flow of copper atoms along the (5,5)–(4,4) CNT channel. Figure 4.6a shows the side and cross-sectional views of the transportation of copper atoms along a (5,5)–(4,4) CNT channel. There are two copper atoms specifically highlighted in dark grey and light grey in Fig. 4.6. The dark and light grey represent the copper atoms numbers 1 and 2, respectively. The placement of the dark and light grey copper atoms in Fig. 4.6b indicates the initial orientation of the copper mass. As the position of the coloured copper atoms changes, the new position represents the change in the orientation of the entire copper mass as shown in Fig. 4.6b and c. At a bias voltage of 2 V, the copper mass is rotated by an angle of approximately 45 degrees with respect to the y coordinate as the copper atoms are propelled towards the (4,4) CNT channel during the simulations.

When copper atoms are translocated along the CNT junction due to the electromigration force, a rotating phenomenon on the copper atoms occurs and causes the potential energy of copper atoms to increase. This rotating process results in a higher total energy for the copper mass, and thus improves the ability of the copper atoms to overcome the energy barrier of the junction.

Figure 4.7 shows the temperature ratio of copper atoms during electromigration transportation to the initial temperature of copper. The temperature ratio for the copper mass shows the effect of an electromigration process on the temperature of the copper mass. The temperature ratio is defined as the ratio of the temperature of copper atoms during electromigration to the temperature of copper atoms at the initial state. When the temperature ratio is high, it is suggested that the energy contributed by electromigration is mainly used to increase the thermal vibration of copper atoms. On the other hand, a low temperature ratio suggests that the energy contributed by electromigration is mainly used to translocate the copper atoms towards the narrow ending of the channel instead. The temperature ratio of the copper mass at a low bias voltage is quite small. The data in Fig. 4.7 suggest that a low bias voltage can in fact only affect the temperature ratio of the copper mass minimally. Since the temperature ratio is small, the thermal vibration of the copper atoms would not change significantly compared to the initial state.

When the bias voltage increases, the temperature ratio of the copper atoms would increase accordingly as well. At 673 K, the maximum temperature ratio for

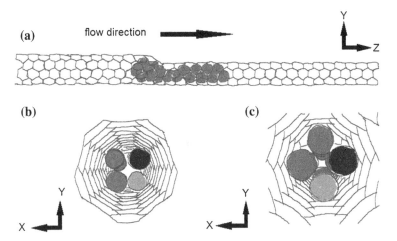

Fig. 4.6 Transportation of copper atoms along a (5,5)–(4,4) CNT channel. (**a**) Side view of copper atoms transported along the (5,5)–(4,4) CNT channel, (**b**) Copper transport along the (5,5) CNT segment and (**c**) Copper atoms transport along the (5,5)–(4,4) CNT junction

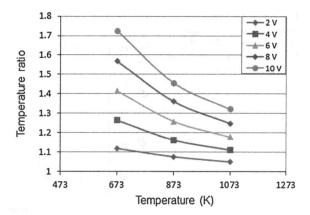

Fig. 4.7 Temperature ratio of the copper mass due to an electromigration versus the initial temperature of the copper mass

a bias voltage of 10 V is about 1.7, which is about 54 % more than the effect of 2 V. As the initial temperature of the copper mass increases, the temperature ratio of the copper mass would also reduce. The reduction in the temperature ratio at high temperatures suggests that the thermal vibration of the copper atoms at high temperatures would be large enough such that any further increase in the bias voltage only contributes minimally to the overall thermal vibration of the copper atoms. This phenomenon would be reasonable assuming that there is no sudden increase in the temperature of copper atoms within a few picoseconds due to electromigration.

Figure 4.8 shows the total interactive forces between the copper mass and a (5,5)–(4,4) CNT channel junction due to different bias voltages. The interactive force is defined using the derivative of Eq. (2.3) with respect to r. The interactive forces between the copper mass and the CNT channel junction are high when the applied bias voltages are high. As mentioned earlier, the copper atoms adopt a rotating phenomenon while being transported along the channel junction at a lower bias voltage. By comparing this phenomenon with Fig. 4.8, it is understood that the rotating phenomenon at a lower bias voltage results in less repulsive interaction between copper atoms and the CNT channel junction because lower interactive forces suggests more attraction between copper atoms and the CNT channel junction.

When the bias voltage is increased, the interaction between copper atoms and the CNT channel junction becomes stronger because there is more repulsion between copper atoms and the CNT junction. According to Fig. 4.8, the highest interactive forces occur when the temperature of the copper atoms is initially set at 673 K, with a bias voltage of 2 V. For initial temperatures that are higher, the interactive forces do not increase as much as that at 673 K. The interactive forces at high bias voltages are suggested as having more repulsion on the copper atoms due to the blockage of the CNT channel junction.

Figure 4.9 shows the changes in the interactive forces due to different initial temperatures of copper atoms in the (5,5)–(4,4) CNT channel. The interactive forces between copper atoms and the CNT junction are higher when initial

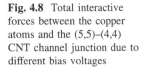

Fig. 4.8 Total interactive
forces between the copper
atoms and the (5,5)–(4,4)
CNT channel junction due to
different bias voltages

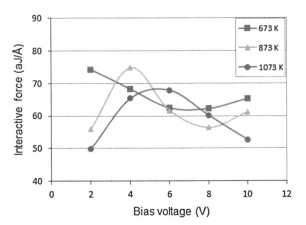

temperatures are low, as compared to that at higher temperatures. The change in
the interactive forces is reflected as the change in the thermal vibration of copper
atoms. As temperature rises, the thermal vibration of copper atoms increases. The
high vibration level causes copper atoms to interact more frequently with the
channel junction, and the attraction between copper atoms and the CNT junction is
reflected as lower interactive forces compared to that at lower initial temperatures.
The copper atoms are less mobile and more compact at lower temperature, and
hence the lesser attraction between copper atoms and the CNT junction is reflected
as higher interactive forces.

Figure 4.10 shows the radial forces of copper atoms translocating along a (5,5)–
(4,4) CNT channel. The direction of the radial forces is defined along the plane
that is parallel to the cross-sectional area of the CNT channel. The CNT channel
junction is represented between −11 and 2 Å in Fig. 4.10. According to Fig. 4.10,
the most significant changes in the radial forces occurs when copper atoms are
transported along the channel junction, as shown between −13 and 0 Å. Apart

Fig. 4.9 Total interactive
forces between copper atoms
and a (5,5)–(4,4) CNT
channel junction due to
different initial temperatures

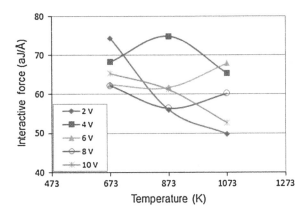

Fig. 4.10 Radial forces on the translocated copper atoms along a (5,5)–(4,4) CNT channel. The initial temperatures are set at (**a**) 673 K, (**b**) 873 K, (**c**) 1073 K respectively

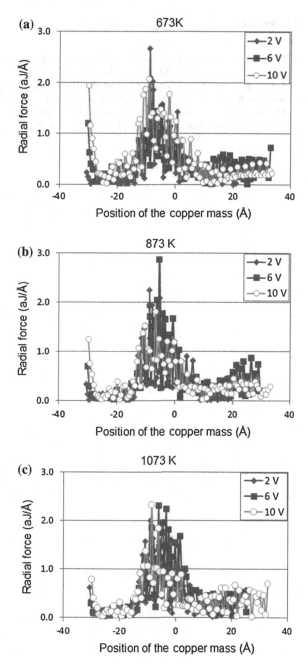

from that, the radial forces in the (5,5) CNT segments are relatively lower, and the radial forces in the (4,4) CNT segment show a slight oscillation while copper atoms move towards the tube end.

References

1. R. Saito, G. Dresselhaus, M.S. Dresselhaus, Tunneling conductance of connected carbon nanotubes. Phys. Rev. B. **53**, 2044–2050 (1996)
2. J. Cai, Y.Y. Ye, Simple analytical embedded-atom-potential model including a long-range force for fcc metal and their alloys. Phys. Rev. B. **54**, 8398–8410 (1996)
3. J.W. Kang, Q. Jiang, H.J. Hwang, A double-walled carbon nanotube oscillator encapsulating a copper nanowire. Nanotechnology **17**, 5485–5490 (2006)

Chapter 5
Variation in Carbon Nanotube Channels as Nanodelivery Systems

5.1 Introduction

This section investigates a variety of CNT channels as nanodelivery systems by varying the sizes of the channels. MD simulations are performed to analyse the effect of the CNT channel junctions on the flow of copper atoms due to electro-migration, and the effect of chirality on the flow behaviour of copper atoms.

The variation in the CNT channel junction sizes is created by combining CNTs of different diameters. The joining of two CNTs with different diameters results not only in dissimilar input and output diameters of the junction, but also inner surface areas of various sizes, which is crucial when the reconfiguration of the copper atoms takes place at the junction. The simulation conditions in this section will be focused on providing the details and parameters for setting up the case studies, and the results and analyses from the simulations will be presented following that.

5.2 Effects of Junction Sizes on the Flow of Copper Atoms Along CNT Channels

In this section, the CNT channels are modelled using nonpolarised, single-walled and armchair CNTs. Each CNT channel consists of one CNT junction, which is joined by two CNTs of different diameters. There are four CNT channels constructed for this simulation study: the (12,12)–(5,5) CNT channel, the (12,12)–(7,7) CNT channel, the (12,12)–(9,9) CNT channel and the (12,12)–(10,10) CNT channel. The diameters for the (5,5) nanotube, (7,7) nanotube, (9,9) nanotube, (10,10) nanotube and (12,12) nanotube are 6.77 Å, 9.48 Å, 12.2 Å, 13.5 Å and 16.2 Å, respectively. The CNT channels span the z coordinate between -65 and 105 Å, with lengths of about 110 and 170 Å. Both the energy minimisation and

M. C. G. Lim and Z. W. Zhong, *Carbon Nanotubes as Nanodelivery Systems*, SpringerBriefs in Applied Sciences and Technology, DOI: 10.1007/978-981-4451-39-0_5, © The Author(s) 2013

equilibrium MD runs are made before the nonequilibrium MD simulations are performed.

The main purpose of the CNT channel junction in this study is to provide a geometrical constraint along the channel for the flow of copper atoms. The carbon atoms are fixed in this study so that the simulation cases are simplified, and the geometrical obstruction along the channel is focused upon instead of considering the effect of moving carbon atoms. Before the simulations are performed, copper atoms are placed near the opening of the (12,12) CNT for all the simulation cases, and the copper mass is aligned parallel to the tube axis. Each CNT channel has 100 copper atoms. The EAM potential in Eq. (2.1) is used to model the copper atoms with a cut-off distance of $1.65a_0$ ($a_0 = 3.615$ Å) for each pair of copper atoms. The van der Waals energy between a pair of copper atom and carbon atom is described using Eq. (2.3).

In order to induce a driving force for the copper atoms to be transported along the CNT channel, electromigration is applied along the channel with a bias voltage of 10 V. The intensity of the electromigration is controlled using Eq. (2.4). The repulsive effect between atoms is represented by the positive values of energies or forces generated from the simulation results, whereas the attractive effect between atoms is represented by the negative values of the simulation results.

The modified velocity rescaling method, shown in Eq. (4.1), is used to control the temperature of the copper atoms. The neighbour list is constructed in all the simulation cases in order to improve the efficiency of the simulation program.

The temperature for the simulation studies in this section is set at 1073 K, which is below the melting temperature of copper. The simulation time step is set at 1 fs. The total simulation duration for the case studies in this section is set at 20 ps, which is sufficient for the copper atoms to be transported from one end of the channel to the other.

5.3 Analyses of Copper Due to Different Junction Sizes

Figure 5.1 shows the transportation of copper in different CNT channels. There are four different cases in this study, whereby each case focuses on one type of CNT channels. The (12,12)–(5,5) CNT channel, (12,12)–(7,7) CNT channel, (12,12)–(9,9) CNT channel and the (12,12)–(10,10) CNT channel are described in cases 1, 2, 3 and 4 respectively. There are a total of 2048 carbon atoms in the (12,12)–(5,5) CNT channel, 1854 carbon atoms in the (12,12)–(7,7) CNT channel, 1956 carbon atoms in the (12,12)–(9,9) CNT channel, and 1998 carbon atoms in the (12,12)–(10,10) CNT channel. The (12,12) CNT segment of each channel is about 45 Å in length, whereas the (5,5) CNT segment is about 100 Å in length, and the (7,7), (9,9) and (10,10) CNT segments are about 60 Å in length.

The (12,12) CNT is defined as the upstream section of the channel, and the (5,5) CNT, the (7,7) CNT, the (9,9) CNT and the (10,10) CNT are the downstream section of the channel. Since the diameter of the narrow section of the CNT

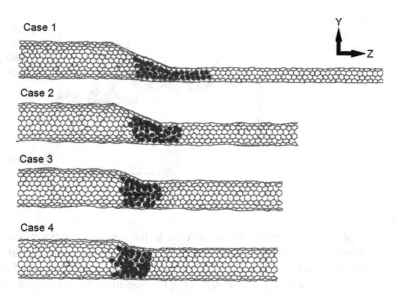

Fig. 5.1 Transportation of copper atoms along 4 different CNT channels. Cases 1, 2, 3 and 4 refer to the (12,12)–(5,5) CNT channel, the (12,12)–(7,7) CNT channel, the (12,12)–(9,9) CNT channel and the (12,12)–(10,10) CNT channel, respectively

channels is different in each case, the formation of the CNT junctions in terms of the length and the inner surface areas is hence dissimilar for each case, which affects the overall length of the channel. In this section, the "entrance" of the channel refers to the position where the (12,12) CNT segment and the channel junction are joined. The "exit" of the channel refers to the position where the downstream segment and the channel junction are joined. The diameter ratio is defined as the quotient of the cross-sectional diameters of the upstream CNT and the downstream CNT. The diameter ratio is also the quotient of the cross-sectional diameters of the channel junction entrance and the exit. The calculated diameter ratios for cases 1, 2, 3 and 4 are 2.4, 1.7, 1.3 and 1.2, respectively. In order for the comparison between different CNT channels to be relevant, the copper atoms are initially located at the same position for each of the CNT channels with the same amount of copper atoms. The results from the simulations are presented in Figs. 5.2, 5.3, 5.4, 5.5, 5.6, 5.7.

When the copper atoms are transported along the CNT channel, the obstruction from the CNT channel junction causes the copper atoms to undergo atomic reconfiguration. In order to provide more meaning to the reconfiguration of the copper atoms, the potential energy of the copper atoms during simulations is plotted. Figure 5.2 shows the potential energy of copper during the transportation along CNT channels. When copper atoms reach the junction, the obstruction from the junction stops the approaching copper atoms. However, the electromigration force continues to push the copper atoms towards the CNT channel junction which leads to the shortening of the separation between the copper atoms in the copper

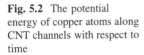

Fig. 5.2 The potential energy of copper atoms along CNT channels with respect to time

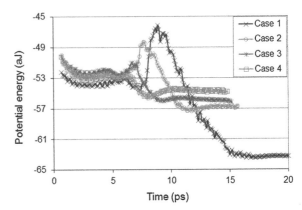

mass. This phenomenon results in an increase in the potential energy of copper atoms, whereby most of the copper atoms are repelling one another and the reconfiguration of copper atoms occurs, as shown in cases 1 and 2 in Fig. 5.2.

Cases 3 and 4 show a very low increment in the potential energy between 5 and 10 ps, compared to cases 1 and 2 in Fig. 5.2. This is because the diameter ratios of the channel junctions for cases 3 and 4 are relatively smaller than the ratios in cases 1 and 2. As the copper atoms reach the exit of the channel junction eventually, the potential energy of copper atoms then rises to the maximum. Although the kinetic energy of copper atoms is reduced due to the decrease in the velocities at the channel junction, the increase in the potential energy of copper atoms and the electromigration forces are sufficient enough to overcome the potential energy barrier at the channel junction caused by the carbon atoms.

The potential energy barrier of the channel junction is closely related to the diameter ratio of the channel junction. Since the diameter ratio of the channel junction is larger in cases 1 and 2, the increment in the potential energy for copper atoms in these two cases is also much more significant than other cases. Based on the results of the simulations, it is noted that a difference of 16.67 % between the upstream and the downstream cross-sectional diameters results in a nearly insignificant increase in the potential energy of copper atoms at the junction, as shown in case 4 of Fig. 5.2. However, a 58 % difference between the upstream and the downstream cross-sectional diameters causes an increase of 18 % in the potential energy of copper atoms at the channel junction in case 1.

The Van der Waals energy between copper atoms and carbon atoms of the CNT channel in the z direction are shown in Fig. 5.3. The position of the copper mass is determined using the position of the centre of mass of the copper atoms. The interactive forces in the 4 cases are compared to illustrate the effect of the junction size on the impedance of the copper flow. The junction of the CNT channel is tapered along the z direction of the channel, which gives a direct obstruction towards the flow of copper atoms. Figure 5.3 shows that the strongest van der Waals force between copper atoms and the carbon atoms of the CNT channel occurs in case 1. The van der Waals forces (between -20 and 20 Å in Fig. 5.3) in

Fig. 5.3 Van der Waals forces between copper atoms and carbon atoms of the CNT channels during simulation

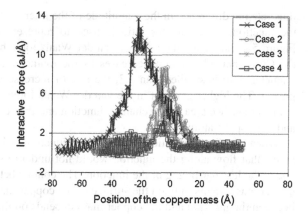

cases 2, 3 and 4 are relatively smaller compared to case 1. The small increase in the interactive forces (between −20 and 20 Å in Fig. 5.3) suggests that the channel junction plays a major role in increasing the van der Waals forces between copper atoms and the carbon atoms of the CNT channel. The upstream and downstream segments of the channels that are straight have a nearly negligible effect on the interactive forces in the z direction.

The magnitude of the interactive force in case 1 shows a gradual reduction over a distance of 20 Å after the 0 Å position, as shown in Fig. 5.3. Since the cross-sectional area of the downstream segment is small, the length required for accommodating the copper atoms at the downstream segment would be longer too. The gradual reduction in the interactive forces suggests that there are remnants of copper atoms in the junction as the rest of the copper atoms flow into the downstream segment of the channel. The ease of leaving the channel junctions, such as in cases 3 and 4, would result in a shorter interval for the interactive forces to reduce to nearly zero.

In order to obtain a better understanding of the relationship between the average interactive forces and the diameter ratio of the channels, Fig. 5.4 is plotted to

Fig. 5.4 Average forces between the copper mass and the channel junction with respect to different diameter ratios of the channels

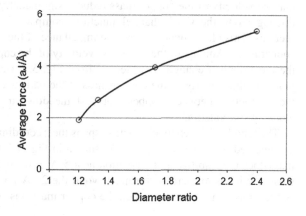

illustrate the difference in the magnitude of the interaction in regard to the sizes of the junction. The diameter ratio is used to represent the size of the junction. Figure 5.4 shows that the average van der Waals force between copper atoms and the CNT channel junction decreases as the diameter ratio increases. When the diameter ratio is smaller than 1.3, the rate of increase in van der Waals force is greater. The higher gradient in the van der Waals force can be explained based on the inner surface area of the channel junction and the contact between the junction and the copper atoms.

When the diameter ratio of the channel junction is lower than 1.3, the copper atoms that flow along the junction would not undergo significant reconfiguration due to the less narrowing of the junction. The contact between the junction and the copper mass occurs only on the surface of the copper mass. The amount of contact between the junction and the copper mass depends on the inner surface area of the junction. The inner surface area would be much larger for the channel junction that is more tapered (diameter ratio increased) towards the downstream segment of the channel due to the difference in the diameters of the upstream and the downstream segments. As the diameter ratio increases, the contact between the inner surface of the junction and the surface of the copper mass would also increase, and this is reflected as an increase in the average interactive force. However, as the diameter ratio continues to increase further, the narrowing of the downstream segment of the channel causes the atomic structure of the copper mass to be rearranged so that the copper atoms can be transported through the exit of the junction by electromigration.

The rearrangement of the copper atoms results in the increase in the surface area of the copper mass, mainly due to the change in the volume of the copper mass. Since both the inner surface area of the junction and the surface area of the copper mass increase, the resulting proportional increment in the average interactive force becomes steady and less steep. According to Fig. 5.4, the low gradient of the van der Waals force occurs at diameter ratios higher than 1.3.

The velocity of the copper mass along the CNT channel is illustrated in Fig. 5.5. The position of the copper mass is determined based on the centre of mass of the copper atoms along the channel in the z direction. Figure 5.5 shows that the velocity of the copper mass reduces significantly when the copper mass is flowing along the CNT channel junction, compared to the velocity at straight sections. This phenomenon shows the impedance of the junction on the flow of the copper mass. Among all the cases, the velocity of the copper mass in case 1 shows the most distinct change, as observed from Fig. 5.5. Besides the influence of the junction, the velocity of the copper mass at the narrow channel is also affected by the interaction between copper atoms at the downstream CNT and the copper atoms at the junction.

The length of the channel junction spans the z coordinate between -21 and 5 Å, as indicated in case 1 of Fig. 5.5. As shown in Fig. 5.5, the velocity of the copper mass does not stabilize at the position near 5 Å. Instead, the velocity of the copper mass continues to increase further beyond the 20 Å position. The velocity of the copper mass stabilizes only when the copper mass has completely passed through

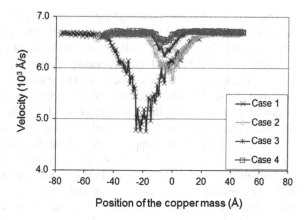

Fig. 5.5 Velocities of the copper mass along the CNT channel during simulations

the junction. The increase in the velocity of the copper mass at the 20 Å position and beyond signifies the existing interaction between the copper atoms at the junction and at the downstream CNT. As the cross-sectional diameter of the downstream CNT channel in cases 2, 3 and 4 increases as compared to case 1, the copper atoms are able to flow through the junction easily and the change in the velocity at the junction is hence smaller.

Another interesting finding in relation to the flow of the copper atoms along the channel is that the difference between the velocity of copper atoms in the upstream CNT and the downstream CNT is rather small. Figure 5.6 shows the velocity ratio of copper atoms with respect to the diameter ratio of the channel. The velocity ratio is defined as the ratio of the velocity of copper atoms at the upstream CNT channel to the velocity of copper atoms at the downstream CNT channel. Likewise, the diameter ratio is defined as the ratio of the cross-sectional diameter of the upstream CNT channel to the cross-sectional diameter of the downstream CNT channel. Figure 5.6 shows that as the copper mass moves from the upstream to the downstream CNT channel, the resulting velocity of the copper mass in the downstream (12,12)–(5,5) CNT channel does not increase, but decreases slightly. The velocity ratio for this phenomenon is reflected as more than 1.

Fig. 5.6 Velocity ratios of the copper mass with respect to the different diameter ratios of the CNT channel

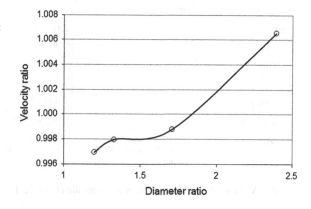

The velocity of copper atoms along the downstream CNT for case 1 is reduced because the flow is faced by the resistance of the inner surface of the downstream CNT channel. The van der Waals interaction between copper atoms and the CNT junction is higher for cases 1 and 2, because the reconfiguration of the copper atoms leads to the elongation of the copper mass in the narrow downstream CNT channel, as shown in Fig. 5.3. The elongation of the copper mass in cases 1 and 2 leads to a larger contact area with the downstream CNT channel, which results in higher resistive forces acting on the surface of the copper mass, especially for case 1.

The flow velocity of copper atoms along the downstream segment of the channel increases for cases 2, 3 and 4, except in case 1. However, the magnitude of the increment is still quite minimal. In this simulation study, one of the significant difference between the classical theory and the atomic simulation model is that in the latter model, the movements of the flowing atoms and the interaction of these atoms with the atoms of the channel are being taken into considerations. The results obtained from the simulations for these atoms are therefore a time average of the flow behaviour for the flowing atoms, rather than an outcome for flow materials as a whole.

Figure 5.7 shows the velocity distribution of copper atoms along the CNT channels. The main difference between Figs. 5.5 and 5.7 is that the former figure shows the average velocity of the copper mass as a whole, whereas the latter shows the z velocity of every copper atom along the CNT channels. The velocity scale consists of both positive and negative values. The positive values refer to the

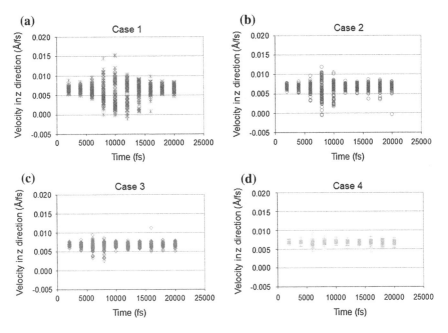

Fig. 5.7 Velocity distributions of copper atoms along the CNT channels

direction of copper atoms moving towards the downstream, whereas the negative values refer to the direction of copper atoms moving towards the upstream.

Among the four cases shown in Fig. 5.7, the velocities of copper atoms at the junction of (12,12)–(5,5) and (12,12)–(7,7) CNT channels (Cases 1 and 2 respectively) vary the most, at 10000 ± 500 (fs). The large variation in the magnitude of the velocities is due to the jamming of copper atoms at the junction of the channel. The velocities of the trailing copper atoms decrease as a result of the blockage by the junction. The squeezing effect of the copper mass at the junction causes the leading copper atoms to be forced out from the junction into the downstream. The increase in the velocities of the leading copper atoms at 10000 ± 500 (fs), as shown in Fig. 5.7, suggests that the leading copper atoms are forced out from the junction due to electromigration. On the other hand, the wide variation in the velocities of copper atoms is noticeably less significant for both the (12,12)–(9,9) and (12,12)–(10,10) CNT channels, which are Cases 3 and 4 respectively.

Chapter 6
Carbon Nanotube Chirality and the Flow Phenomena of Copper Atoms

6.1 Introduction

The last chapter of this book focuses on the investigation of the effect of the CNT chirality on the flow behaviour of copper atoms. The variation in the CNT channels is created using CNTs with different chirality. The purpose of creating the CNT channels with different chirality is to use these nanodelivery systems to find out how the distribution of carbon atoms along the channels would influence the flow behaviour of nanomaterials, particularly for copper atoms in this section of the book.

The investigation is carried out using similar MD simulation methods as that described in previous chapters. As mentioned in Chap. 4, the copper atoms displayed a spiral flow phenomenon when copper atoms are being transported along a (5,5)–(4,4) CNT channel. Based on this particular finding, the investigation of copper flow is extended for another type of CNT channels of similar sizes. The two main characteristics of CNTs, which are chosen for this study, are the armchair and zigzag CNTs.

6.2 Effect of CNT Chirality on the Spiral Flow of Copper Atoms in Their Cores

MD simulations are performed in this chapter for the flow of copper atoms along CNT channels with different chirality. In the simulations, the two CNT channels are modelled using nonpolarised, single-walled, zigzag and armchair CNTs. Each of the CNT channels consists of one CNT junction, formed by the joint of two nanotubes of different diameters. The (5,5)–(4,4) CNT channel consists of two

M. C. G. Lim and Z. W. Zhong, *Carbon Nanotubes as Nanodelivery Systems*,
SpringerBriefs in Applied Sciences and Technology,
DOI: 10.1007/978-981-4451-39-0_6, © The Author(s) 2013

armchair CNTs, whereas the (8,0)–(7,0) CNT channel consists of two zigzag CNTs.

In order to compare the effect of the CNT chirality on the flow of copper atoms, the CNT channels are chosen such that there are minimal differences in the diameters between the two channels. This is to ensure that the comparison is not significantly affected by the geometrical differences between the two CNT channels. There are 828 carbon atoms in the (5,5)–(4,4) CNT channel, and the diameters of the (5,5) nanotube and the (4,4) nanotube are 6.8 Å and 5.4 Å, respectively. There are 795 carbon atoms in the (8,0)–(7,0) CNT channel, and the diameters of the (8,0) nanotube and the (7,0) nanotube are 6.3 Å and 5.5 Å, respectively. The two CNT channels span the z coordinate between −53 and 59 Å, which is approximately 112 Å in length. The carbon atoms of the CNT channels are fixed in space.

Each of the CNT channels consists of 28 copper atoms. The copper atoms are located at the widening ends of the CNT channels. Energy minimisation and equilibrium MD runs are made before the nonequilibrium MD simulations are carried out.

The copper atoms are modelled with the EAM potential shown in Eq. (2.1). The interaction between the copper atoms and the carbon atoms is described using the LJ potential in Eq. (2.3), whereas Eq. (2.4) describes the electromigration force necessary for driving the copper atoms along the CNT channels. In order to control the temperature of the system during simulations, the modified velocity rescaling method in Eq. (4.1) is used to adjust the temperature.

There are two bias voltages set in this simulation study: 2 and 4 V. 2 V is chosen as one of the simulation conditions so that the erratic behaviour of copper atoms due to the strong driving forces can be minimised. The repulsive effect between two atoms in the simulations is described by the positive values of energies or forces, whereas the attractive effect is described by the negative values instead.

There are two initial temperatures set for the simulation cases: 673 and 1073 K. The simulation cases are set up based on the combinations of temperatures, orientation of the copper mass and the types of CNT channels used, which will be elaborated in the next section. The time step for the simulations is set at 1 fs. The maximum simulation duration for this study is set at 20 ps, which is necessary for the transportation of copper atoms to be completed along the CNT channel.

In order to simplify the case studies, the force field parameters for the carbon atoms in the CNT channels are not established. Instead, the focus on the difference between the CNT channels is solely on the spatial distribution of carbon atoms along the channel. The electronic properties of CNTs associated with the different chirality will also be neglected for this study to reduce the factors influencing the flow phenomena of copper atoms. The main contribution for the difference in the flow behaviour of copper atoms therefore depends on the LJ potential between copper and carbon atoms at this stage.

6.3 Analyses of the Spiral Flow Phenomenon of Copper Atoms

The effect of the CNT chirality on the spiral flow phenomenon of copper atoms is investigated in this section. The driving force for the flow of copper atoms is achieved through the induction of electrmogration along the channel. The copper atoms flow from the wider end (upstream) of the CNT channel towards the narrow end (downstream) of the CNT channel. The magnitude of the driving force depends on the amount of the bias voltage applied along the channel.

Figure 6.1a shows the translocation of copper atoms along a (5,5)–(4,4) CNT channel (armchair CNT channel) with a bias voltage of 4 V at 673 K. The cross-sectional view of the initial configuration of the copper atoms at the wider end (5,5) CNT, is shown in Fig. 6.1b. The copper atoms depicted in Figs. 6.1b–d are the first four atoms of the copper mass. Since the structure of the copper mass is made up of the repetition of the first four copper atoms, the first four copper atoms are thus used as a representation to show the cross-sectional movements of the copper mass along the CNT channel.

When the copper atoms move along the CNT channel due to electromigration, the structure of the copper mass undergoes reconfiguration as the copper mass passes through the junction of the channel. Due to the narrowing of the diameter along the junction, the copper atoms are forced to undergo rearrangements before the copper mass is able to pass through the channel junction, through the combined effect of the driving force and the interaction between copper and carbon atoms. Such a condition leads to the reconfiguration of the copper mass in the (4,4) CNT as shown in Fig. 6.1c. As more copper atoms continue to flow into the narrow segment of the channel, copper atoms that have left the junction tend to move in a spiral fashion, as observed in Fig. 6.1d. The spiral flow behaviour has been mentioned earlier in Chap. 4. However, a more thorough investigation shows that

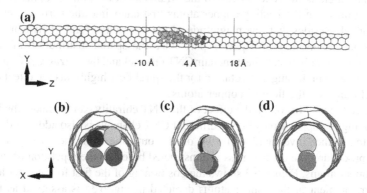

Fig. 6.1   Translocation of copper atoms along a (5,5)–(4,4) CNT channel at 673 K, with a 4 V bias voltage. Sketches show the cross-sectional view of the CNT channel at **a** 5.4 ps **b** 8.1 ps, and **c** 15.1 ps respectively

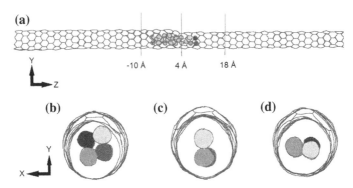

Fig. 6.2 Translocation of copper atoms along a (8,0)–(7,0) CNT channel at 673 K, with a 4 V bias voltage. Sketches show the cross-sectional view of the CNT channel at **a** 5.4 ps **b** 8.1 ps, and **c** 15.1 ps respectively

the copper atoms not only rotate while flowing, but forms a two copper atoms arrangement visible through the cross-sectional view of the (4,4) CNT.

Figure 6.2a shows the flow of copper atoms along a zigzag CNT channel. Although the initial configuration of the copper atoms in the (8,0) CNT, as shown in Fig. 6.2b, is nearly similar to the arrangement of the copper atoms in the (5,5) CNT, the final structure of the first four copper atoms in the narrow section of the (7,0) CNT has turned out to be different from that in the (4,4) CNT. The first four atoms of the copper mass are depicted in Figs. 6.2b–d. Figure 6.2c shows the restructured form of the first four copper atoms in the (7,0) CNT when the copper atoms have just entered into the (4,4) CNT segment.

Figure 6.1c shows that the distribution of the copper atoms is similar to Fig. 6.2c. However, as the copper atoms proceeds farther in the (7,0) CNT segment, the first four copper atoms rotate about 90° to the right. This angle of rotation is much larger than that occurred in the (4,4) CNT. The cause of the larger rotational angle is likely to be due to the arrangement of trailing copper atoms at the junction when the leading copper atoms first enter into the narrow section of the channel. Besides that, the magnitude of the interactive forces between copper and carbon atoms may also play a part since the spatial distribution of carbon atoms is different between the armchair CNT channel and the zigzag CNT channel. The difference in the angle of rotation for the spiral flow highlights the effect of the CNT chirality on the flow of copper atoms.

Before the conclusion on the effect of the CNT chirality can be made, the initial positioning of copper atoms in two different CNT channels is also addressed in this study for further investigation. Based on the outcome from the energy minimisation procedure, the copper mass is constructed based on the repetition of the first four copper atoms. Figure 6.3 shows the positioning of the first four copper atoms. The arrangement of the copper atoms depicted in Fig. 6.3a is assigned as the 0° configuration for this study. Since the CNT junctions of the channels are asymmetrical, another configuration for the placement of copper atoms is also

Fig. 6.3 Initial configuration of the first four copper atoms with **a** 0° of rotation, and **b** 45° of rotation

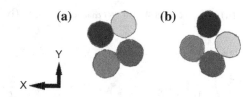

considered, and is shown in Fig. 6.3b. The placement of the copper atoms in Fig. 6.3b is assigned as the 45° configuration. The reason for having two different initial configurations of the copper atoms is such that the interaction between the copper mass and the CNT channel junction can be different based on the initial contact between these two bodies.

The factors that may influence the spiral flow phenomenon of the copper mass include temperature, the initial positioning of the copper atoms, the magnitude of the bias voltages and the interaction of carbon atoms in CNT channels with the encapsulated metal atoms. From Figs. 6.4, 6.5, 6.6, 6.7, each of the possible parameters that may affect the outcome of the flow behaviour is investigated in this chapter, so as to verify that the chirality of the CNT channel is the main contributing factor.

Figure 6.4 shows the cross-sectional view of the spatial distribution of the first four copper atoms along the CNT channel with a 2 V bias voltage at 673 K. The copper atoms converge to the centre of the channel as they move downstream. The copper atoms are reconfigured from four positions into two positions, as depicted in the cross-sectional views of the channels in Fig. 6.4. A similar repositioning of the copper atoms is also observed in Figs. 6.1c and 6.2c. The central position of the downstream CNT channel is between 0 Å and −1 Å for the y axis and 0 Å for the x axis due to the asymmetric feature of the downstream segment of the channel. The copper atoms converge towards the centre of the channels regardless of the initial position of the copper atoms. Since three out of the four copper atoms in the 45° configuration are positioned closer to the central axis of the downstream segment of the channel (between 0 Å and −1 Å for the y axis), the cross-sectional path taken by the copper atoms in this configuration is also shorter. The spiral flow phenomenon in this case is therefore quite insignificant. As compared between figures a–d in Fig. 6.4, it is observed that the spiral flow phenomenon in the (8,0)–(7,0) CNT channel is more obvious than that in the (5,5)–(4,4) CNT channel when the copper atoms are at the 0° configuration. It is also noted that the low bias voltage (2 V) results in a weaker driving force, and hence the repulsion by the junction of the channel onto the copper atoms is not strong. The convergence process of the copper atoms along the downstream is therefore smoother.

Figure 6.5 shows the flow of the first four copper atoms along the channels at 1073 K, with a 2 V bias voltage. The increase in temperature, at 1073 K, causes a greater vibration among the copper atoms, compared to that in Fig. 6.4, at 673 K. Since the high temperature increases the thermal vibration of copper atoms, the random motions of the copper atoms do not form a significant spiral flow fashion when they are transported along the downstream of the CNT channel.

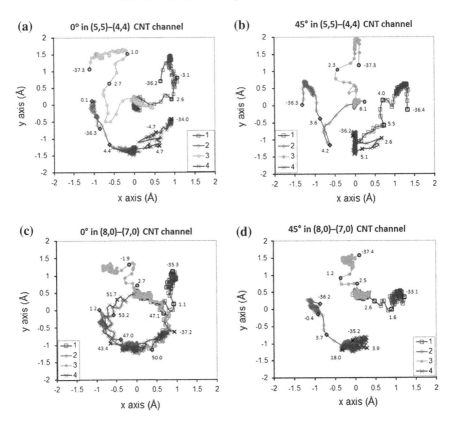

Fig. 6.4 The cross-sectional view of the spatial distribution of the first four copper atoms along the CNT channel with a 2 V bias voltage at 673 K. The first four copper atoms at **a** 0° in the (5,5)–(4,4) CNT channel **b** 45° in the (5,5)–(4,4) CNT channel **c** 0° in the (8,0)–(7,0) CNT channel, and **d** 45° in the (8,0)–(7,0) CNT channel. The values shown beside the data points represent the z positions of the copper atoms along the CNT channel

The copper atoms are found to be more constrained within the downstream (8,0)–(7,0) CNT channel, than that in the (5,5)–(4,4) CNT channel, as shown in Fig. 6.5. The formation of the spiral flow by copper atoms depends not only on the temperature, but also the distribution of the carbon atoms along different CNT channels, whereby the point of interaction between carbon and copper atoms would vary between the two channels. The resulting structure of the copper mass in the downstream segment of the channel at a higher temperature is similar to a thread, where atoms are connected one after the other along the downstream. This outcome is observable in both the (5,5)–(4,4) and the (8,0)–(7,0) CNT channels.

In order to understand the influence of a higher bias voltage on the spiral flow phenomenon of copper atoms, the flow phenomenon of copper atoms at a bias voltage of 4 V is investigated. Figure 6.6 shows the cross-sectional view of the spatial distribution of the first four copper atoms along the CNT channel with a 4 V bias voltage at 673 K. As the magnitude of the bias voltage increases, the

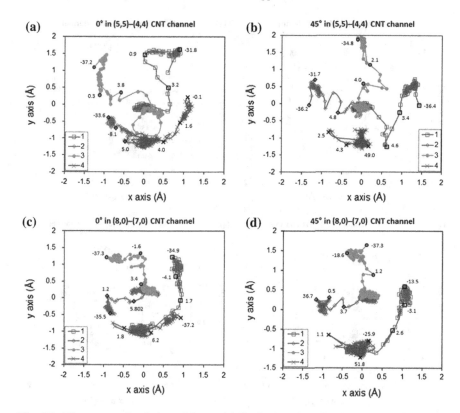

Fig. 6.5 The cross-sectional view of the spatial distribution of the first four copper atoms along the CNT channel with a 2 V bias voltage at 1073 K. The first four copper atoms at **a** $0°$ in the (5,5)–(4,4) CNT channel **b** $45°$ in the (5,5)–(4,4) CNT channel **c** $0°$ in the (8,0)–(7,0) CNT channel, and **d** $45°$ in the (8,0)–(7,0) CNT channel. The values shown beside the data points represent the z positions of the copper atoms along the CNT channel

resulting collision of the copper atoms and the CNT channel junction becomes stronger. The magnitude of the flow of copper atoms to overcome the blockage of the junction so as to proceed farther into the narrow channel becomes more significant than that at a lower bias voltage. As shown in Fig. 6.6, the spiral flow phenomenon of the first four copper atoms is more significant along the (8,0)–(7,0) CNT channel, than that in the (5,5)–(4,4) CNT channel.

Due to the asymmetrical feature of the channel junction, copper atoms at the $45°$ configuration are closer to the central axis of the downstream segment of the channel. The path taken by the copper atoms to move into the junction is thus shorter and the spiral flow phenomenon is less obvious. Apart from that, it is found that the (8,0)–(7,0) CNT channel causes the formation of the spiral flow path by the copper atoms to be more significant than the (5,5)–(4,4) CNT channel. The spiral flow phenomenon by copper atoms is most significant at the $0°$ configuration in the (8,0)–(7,0) CNT channel, as shown in Fig. 6.6c. The reconfiguration of the

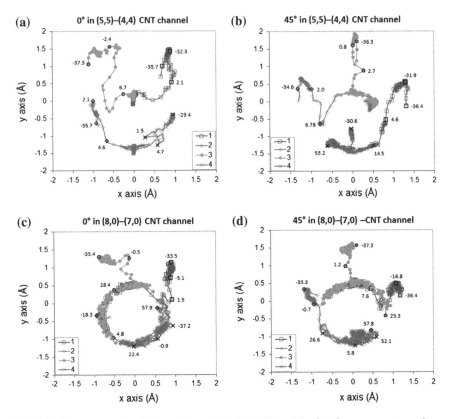

Fig. 6.6 The cross-sectional view of the spatial distribution of the first four copper atoms along the CNT channel with a 4 V bias voltage at 673 K. The first four copper atoms at **a** 0° in the (5,5)–(4,4) CNT channel **b** 45° in the (5,5)–(4,4) CNT channel **c** 0° in the (8,0)–(7,0) CNT channel, and **d** 45° in the (8,0)–(7,0) CNT channel. The values shown beside the data points represent the z positions of the copper atoms along the CNT channel

copper atoms along the CNT channel is a necessary process for the copper atoms to increase its potential energy, so that the potential barrier of the channel junction can be overcome.

The fluctuation of the copper atoms becomes greater when the temperature is increased. Figure 6.7 shows the movements of the copper atoms with a 4 V bias voltage at 1073 K. As compared to Fig. 6.6, it is found that the spiral flow phenomenon of the copper atoms in Fig. 6.7 is significantly reduced and the flow path along the x and y axes is also much shorter. Instead of fluctuating along the x and y planes, the copper atoms move along the CNT channels much quickly, due to the combination of a stronger driving force and a higher temperature. The combined energies from the high temperature and the high driving force on the copper atoms cause the copper atoms to overcome the potential barrier much easily, and flow through the channel junction without much fluctuation, as compared to Fig. 6.5 at the same temperature. The resulting reconfiguration process of the copper atoms is

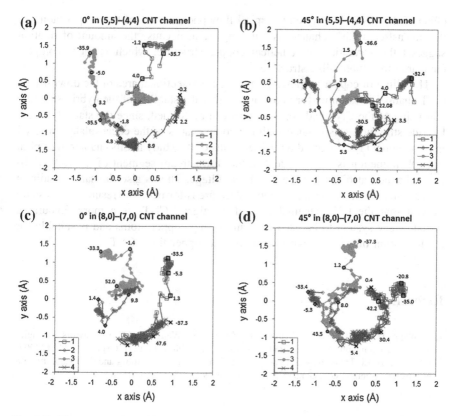

Fig. 6.7 The cross-sectional view of the spatial distribution of the first four copper atoms along the CNT channel with a 4 V bias voltage at 1073 K. The first four copper atoms at **a** 0° in the (5,5)–(4,4) CNT channel **b** 45° in the (5,5)–(4,4) CNT channel **c** 0° in the (8,0)–(7,0) CNT channel, and **d** 45° in the (8,0)–(7,0) CNT channel. The values shown beside the data points represent the z positions of the copper atoms along the CNT channel

thus shorter. Nonetheless, the spiral flow phenomenon is still evident among the copper atoms in Fig. 6.7c.

Despite the similar sizes between the two CNT channels, the outcomes in the flow path are quite different. As shown in Figs. 6.6 and 6.7, the spiral flow phenomenon of copper atoms along the (8,0)–(7,0) CNT channel, at a 4 V bias voltage, is evident. These results suggest that the different chirality of the CNT channel does affect the flow pattern of the copper atoms along the channel. With a different carbon distribution around the CNT channel, the (8,0)–(7,0) CNT channel junction tends to cause the copper atoms to interact more strongly around the inner wall of the channel, which results in a rotational flow path in the downstream segment of the channel.

The spiral flow phenomenon of copper atoms in the CNT channel also results in the formation of a helical structure for the copper mass in the downstream CNT channel. The copper atoms form rows of atoms, which are wounded helically. The effect of the CNT chirality on the spiral flow phenomenon of the copper atoms in the downstream CNT channel is thus important, because the amount of rotation suggests the formation of a helical copper structure, which is more compact compared to a thread-like structure.

The filling of copper as viewed from the cross-sectional area of the downstream CNT channel is essential for knowing whether the stability of the hybrid structure consisting of the CNT and copper atoms can be formed, since the stability of the hydride structure is associated with the compactness of the encapsulated material in the channel [1, 2]. According to the results of the study in this chapter, the spiral flow phenomenon is most evident in the downstream segment of the (8,0)–(7,0) CNT channel, at 673 K with a 4 V bias voltage. This is also the same condition under which the most compact copper structure is formed. The results of this study suggest that the CNT channel formed by zigzag CNTs is a more favourable nanodelivery system to be used for transporting copper atoms and is suitable for the formation of a better hybrid system of a copper filled CNT.

References

1. L. Wang, H.W. Zhang, Z.Q. Zhang, Y.G. Zheng, J.B. Wang, Buckling behaviors of single-walled carbon nanotubes filled with metal atoms. Appl. Phys. Lett. **91**, 051122–051123 (2007)
2. F.W. Sun, H. Li, Torsional strain energy evolution of carbon nanotubes and their stability with encapsulated helical copper nano wires. Carbon **49**, 1408–1415 (2011)

Index

M. C. G. Lim and Z. W. Zhong, *Carbon Nanotubes as Nanodelivery Systems*,
SpringerBriefs in Applied Sciences and Technology,
DOI: 10.1007/978-981-4451-39-0, © The Author(s) 2013